MARINE TOURISM

The sea offers many opportunities for recreation and tourism purposes and in practical terms it is a source of food and transport. In the past, most of our marine environment was 'protected' from tourist use by its inaccessibility, safety concerns and the relatively high cost of recreating in the sea. Through recent decades, significant advances in technology and the increase in international travel have made marine environments more accessible in both real and economic terms. Coastal and marine tourism has become a huge business that forms a significant component of the growing global tourism industry.

Marine Tourism examines successful and unsuccessful tourism with regard to the marine environment. Providing an overview of the history and development of tourism centred on the marine environment, the author moves on to examine the characteristics of marine tourists and considers the roles of 'vendors' of marine tourism activities and opportunities. The final section examines the impacts of tourist activities on marine ecosystems and coastal-based communities and explores management techniques which may be appropriate in reducing negative impacts and maximising the benefits of marine tourism.

Mark Orams is Lecturer in the Centre for Tourism Research at Massey University, Albany, New Zealand.

MARINE TOURISM

Development, impacts and management

Mark Orams

London and New York

First published 1999
by Routledge
11 New Fetter Lane, London EC4P 4EE

Simultaneously published in the USA and Canada
by Routledge
29 West 35th Street, New York, NY 10001

Typeset in Sabon by
Ponting–Green Publishing Services,
Chesham, Buckinghamshire
Printed in Great Britain by
Biddles Ltd, Guildford and King's Lynn

British Library Cataloguing in Publication Data
A catalogue record for this book is available
from the British Library

Library of Congress Cataloging in Publication Data
Orams, Mark
Marine Tourism: development, impacts and management /
Mark Orams.
p. cm.
Includes bibliographical references and index.
1. Tourist trade–Environmental aspects. 2. Tourist trade–Environmental aspects–Case studies. 3. Coastal zone management–Environmental aspects. 4. Coastal zone management–Environmental aspects–Case studies. I. Title.
G155.A1065 1998
333.78–dc21 98-18191
CIP

ISBN 0-415-19572-1 (hbk)
ISBN 0-415-13938-4 (pbk)

For Renée

CONTENTS

ILLUSTRATIONS

PLATES

FIGURES

TABLES

PREFACE

As a preliminary to this book, it is important for me to identify for the reader my paradigm of the marine environment. All writers are inevitably influenced by their experiences and training, and I am no exception. I am unashamedly a marine conservation advocate. It is obvious to me that the future health of our planet and all things that live on it is totally dependent on the sea. That is to say, the health of our world's oceans is critical to the health of our planet. What concerns me, therefore, when looking at the rapid growth of tourism based upon marine resources is the impacts of those tourists' activities and the associated infrastructural developments.

I am strongly influenced by writers such as Sylvia Earle, who expresses the critical importance of our marine environments so well:

> It doesn't matter where on Earth you live, everyone is utterly dependent on the existence of that lovely, living saltwater soup. There's plenty of water in the universe without life, but nowhere is there life without water ...
>
> The living ocean drives planetary chemistry, governs climate and weather, and otherwise provides the cornerstone of the life-support system for all creatures on our planet, from deep sea starfish to desert sagebrush. That's why the ocean matters. If the sea is sick, we'll feel it. If it dies, we die. Our future and the state of the oceans are one.
>
> (Earle, 1995: 30)

Thus, I believe that the basis for analysing and managing marine recreational activities, including tourism, must be ensuring the sustainability of the resource upon which depend, not only the recreation, but the health of all living things. This may seem to be somewhat of an over-reaction. The sea is vast and by far the great majority of marine-based tourism occurs in but a small portion of that vastness. How then can recreational activities in the sea threaten the survival of the planet? This question is addressed in a reflection from one of the best-known of ocean explorers, Jacques Cousteau:

> But soon I had to face the evidence: the blue waters of the open sea appeared to be, most of the time, a discouraging desert. Like deserts on land, it was far from dead, but the live ingredient, plankton, was thinly spread, like haze, barely visible and monotonous. Then, exceptionally, areas turned into meeting places; close to shores and reefs, around floating weeds or wrecks, fish would gather and make a spectacular display of vitality and beauty …
>
> The 'oasis theory' was to help me to understand that the ocean, huge as it may be when measured at human scale, is a very thin layer of water covering most of our planet – a very small world in fact – extremely fragile and at our mercy.
>
> (Cousteau, 1985: 12)

The oceans of our planet are, therefore, not the vast, endless resource that many humans still perceive them to be. Despite the fact that nearly 70 per cent of our planet is covered by ocean, only small portions of this area form the basis for most forms of marine life. I am often reminded of early European pioneering attitudes to land-based resources. Forests, animals (like the North American bison) and minerals were thought to be so plentiful that removing as many of them as we liked or needed would have little impact. We now understand that those attitudes were selfish, short-sighted and wrong. Unfortunately, we now appear to be making the same basic mistakes with marine resources.

Understanding these things is critical if one is to assess adequately the development and impacts of marine tourism and to design or advocate management strategies, as is suggested in the subtitle to this book. My view is, therefore, decidedly conservation oriented, for I cannot see the marine world from any other perspective. My view is also solution oriented, for I care about our oceans and the creatures that live within them; consequently I am motivated to try and find solutions to the negative impacts that are caused by humans and their activities. The fact that my glasses are not rose-coloured but marine blue has undeniably influenced this work. I flag that so that the reader understands this bias. I believe it to be understandable, defensible and even desirable; however, I do recognise that others have a different world view and consequently will have a different slant on this topic.

My hope is that through visiting and enjoying the marine environment as tourists, many people will come to view the oceans as worthy of protection. Perhaps in the same way as tourism is now used as a legitimate justification for the protection of land-based resources (as when such resources are protected in a national park), we will see more marine conservation advocates and marine parks. This hope may be somewhat naive, for the rapid and widespread development of marine-based tourism suggests that it may merely be another form of exploitation of marine resources rather than an agent for marine conservation. However, there are a number of examples where tourism has produced positive results for things marine. It can happen and has happened. This heartens me greatly, for if it can be done once it can be done again and again. This is the challenge that this book seeks to address.

ACKNOWLEDGEMENTS

As is the case with all works of this nature a great number of people have assisted and encouraged me. Staff from the Department of Management and International Business at Massey University Albany have continued to be supportive of research and scholarly writing. In particular my colleagues in the Centre for Tourism Research, Associate Professor Stephen Page, Dr Keith Dewar and Ms Kaye Thorn, have been extremely helpful and have encouraged me a great deal. Stephen has been especially important as a mentor and reviewer of this work; his assistance is much appreciated. My thanks also go to the staff at Routledge for their patience in waiting for the manuscript to be completed and for their help with editing and improving the work.

While marine tourism has not received much specific attention in the literature, I would like to acknowledge a number of important and influential thinkers and writers in this area. Dr Jan Auyong, Professor Mark Miller, Professor Robert Ditton, Dr Paul Forestell, Dr Sylvia Earle, the late Jacques Cousteau and others have provided much-needed inspiration and have shaped my thinking through their work and writings.

I would also like to acknowledge the influence of Sir Peter Blake. Working with Sir Peter has taught me much about determination, organisation and applying oneself to achieve worthwhile things. I thank him for the opportunity to be involved in a number of his 'adventures' over the years.

A number of students from the Centre for Tourism Research have assisted in the production of this book. Thanks go to Annette Lewis for editing and checking references, and particular thanks go to Craig Jones for assistance in obtaining references and compiling information for case studies, and for editorial assistance. Students in an experimental marine tourism course at Florida International University during 1995 also provided useful feedback and ideas during the initial planning for this book.

At a personal level, my parents, John and Lynnette, my brothers, Brett and Simon, and my friends here in New Zealand and around the world continue in their never-ending encouragement and enthusiasm for what I do.

Most of all I must thank my wife, Renée, for her love, for her belief in me and my abilities, and for sharing life with me.

1

INTRODUCTION

Travel for recreational purposes is not a new phenomenon. It has existed for centuries and dates back to pre-biblical times (Adler, 1989). However, it has only become a globally significant enterprise in the latter part of the twentieth century. The growth of tourism has been so dramatic that many claim that it now constitutes the single largest industry in the world (Miller, 1990; Jenner and Smith, 1992).

Tourism has become a significant contributing sector of the global economy. During 1996 the World Tourism Organisation (1997a) estimates that there were a total of 592 million international tourism arrivals, an increase of 4.5 per cent from 1995. Expenditure from these tourists amounted to around $US423 billion, a 7.6 per cent increase on the previous year. Predictions for the future of global tourism are for a continuation of this rapid growth at an average rate of 4.3 per cent per year, to a total of 1.6 billion international travellers in the year 2020 spending more than $US5 billion every day. These predictions estimate that in 2020 there will be three times the number of international tourists there were in 1995 and they will spend nearly five times more (World Tourism Organisation, 1997b). In addition, it must be remembered that many, many more millions of people travel domestically. Tourism is, therefore, a large and rapidly growing global industry. The growth of tourism as an activity, and more recently as an industry, has received much attention in the literature. Organisations such as the United Nations, the World Conservation Union (IUCN), national governments, large corporations, non-profit groups and education institutions, as well as researchers, have all recognised this global phenomenon of travel.

As a result of this interest, a multitude of texts, periodicals and other publications that focus on the travel and tourism industry have been produced since the late 1970s. In recent years a number of authors have begun to examine specialist segments of this large and diverse industry in greater detail. However, few of these works include specific consideration of marine-based tourism. There are, for example, only a handful of papers in periodicals that focus specifically on marine tourism (for example, those by Miller, 1993 and Miller and Auyong, 1991). There are a number that assess the growth and impact of tourism on specific marine locations such as the Great Barrier Reef (Kenchington, 1991; Dinesen, 1995), Korea's coast (Kim and Kim, 1996) and islands of the Indian Ocean (Gabbay, 1986). Some specific marine tourism activities have received attention; for example, a recent issue of the *Journal of Tourism Studies* (volume 7, number 2) focuses on the cruise-ship industry. Probably the most important publication to date on marine tourism is that of the proceedings of the first international conference on the subject, held in Hawaii in 1990 (Miller and Auyong, 1990). The proceedings of a conference held in 1996 have also now been published (Miller and Auyong, 1998). However, there is currently no university-level text that provides a structured consideration of the development, the impacts and the management of marine tourism. While a number of texts consider the management of marine resources (for example, Kenchington, 1990b), marine parks (for example, Salm and Clark, 1989) and coastal zones (for example, Jolliffe, Patman and Smith, 1985), the work contained in this book is the first attempt at an introductory textbook on the subject of marine tourism itself.

WHAT IS THE MARINE TOURISM INDUSTRY?

A diverse range of businesses forms the marine tourism industry. Those directly associated with marine tourism include small, one-person operations such as charter fishing-boat operators, sea-kayak tour guides and scuba-diving instructors. They also include moderate-sized private companies like whale-watch cruise operators and charter-yacht companies, and large, multinational corporations such as cruise-ship companies. An even greater number of businesses and agencies are indirectly associated with marine tourism. Examples include boat maintenance shops, coastal resorts, scuba tank-fill shops, windsurfer rental agencies, fishing equipment suppliers, island ferry services, souvenir collectors, artists and even rubbish collectors. Government agencies also have an important role in the monitoring and management of marine tourism; examples include marine park management authorities, fisheries control agencies, tourism marketing and promotion bodies, law enforcement agencies and marine safety organisations. In addition, there are many non-profit groups that form an important component of the industry, such as clubs for scuba diving, surf life-saving, yachting, windsurfing, surfing and fishing. Environmental groups are also important 'stakeholders' within the industry. Thus, it is both large and diverse.

There is no estimate of the value of the marine tourism industry; however, there is no doubt that the annual turnover is substantial and that it forms a significant business sector with widespread economic influence. This is particularly so for island and coastal communities, where marine tourism can form the single most important economic activity (Miller, 1990).

Because so many businesses, government agencies and other organisations

Plates 1.1 and 1.2 The marine tourism industry is as diverse as an individual windsurfing or people holidaying on a cruise ship.

(such as sport clubs and environmental groups) are involved in marine tourism it is difficult to define the industry tightly. The concept of marine tourism is discussed further and a definition is developed in the next chapter.

WHY STUDY MARINE TOURISM?

Travel to coastal areas for recreation has existed for probably as long as humans have. The sea has a strong attraction for people, which is not surprising given its importance as a source of food and transport. This importance is reflected in the fact that the great majority of the world's population resides along the coast (Miller, 1990). Coastal and marine tourism is, quite simply, a huge business that forms a significant component of the wider tourism industry. For many island and coastal nations it is the primary focus of their tourism industries (Miller and Auyong, 1991). Marine tourism is, therefore, in many situations, the only type of tourism. However, in many cases, it is merely an integral part of the wider tourism industry.

Why then separate marine tourism out from other types? There are several good reasons, first because marine tourism is characterised by different features from other types of tourism. For example, it occurs in an environment in which we do not live and in which we are dependent on equipment to survive. Second, it is growing at a faster rate than most of the rest of the tourism industry. Third, it is having significant negative impacts; and fourth, it presents special management challenges.

'GAIA' AND MARINE TOURISM

The view that our planet is a living organism was originally proposed by James Lovelock in 1979. The hypothesis arose from a study that focused on identifying the conditions that would be necessary to sustain life on other planets. When this same study was applied to our own planet, a new perspective emerged. This was termed 'Gaia', which was defined as:

> A complex entity involving the Earth's biosphere, atmosphere, oceans, and soil, the totality constituting a 'feedback' or 'cybernetic' system which seeks an optimal physical and chemical environment for life on this planet.
>
> (Lovelock, 1985: 100)

Thus, all life, including humans, can be viewed as integral parts of a single living entity, 'Gaia'. From a conservation-oriented paradigm the Gaia hypothesis has value because it recognises the interconnectivity of all living things. More specifically, Gaia recognises that the alteration of one part of an ecosystem will cause changes in another. The following description by Thomas expresses the Gaia idea eloquently:

> The most beautiful object I have ever seen in a photograph, in all my life, is the planet Earth seen from the distance of the moon, hanging there in space, obviously alive. Although it seems at first glance to be made up of innumerable separate species of living things, on closer examination every one of its working parts, including us, is interdependently connected to all the other working parts. It is, to put it one way, the only truly closed ecosystem that any of us knows about. To put it another way, it is an organism.
>
> (Thomas, 1985: 258)

Thomas's 'world view' can be taken a step further. The living organism that is our planet is not 'Planet Earth'; it is more accurately 'Planet Water'. The great majority, around 70 per cent, of our planet's surface is water. These oceans and seas contain a greater diversity of life than terrestrial environments (Earle, 1995). Thus, in terms of Gaia, the functioning of marine environments is enormously influential and important. Furthermore, near-shore environments form the most critical component of the Gaia organism.

Estuaries, wetlands, rocky and coral reefs and protected bays and sheltered harbours are the life-support system of our oceans. This is because the most biologically productive marine areas occur in near-shore environments:

> The narrow coastal fringes of the world's ocean are at once its most productive and most vulnerable zones. Their shallow waters, saturated with sunlight and richly supplied with nutrients, provide the basis of most of our fisheries. Coastal and island ecosystems also serve as a great meeting ground between land and sea; large numbers of people live here, whether in traditional fishing communities or in cities ...
>
> The four vital ecosystems for humankind and for all marine lifeforms are saltmarshes, mangroves, estuaries and coral reefs.
>
> (Myers, 1985: 74)

This fact is what causes the most concern with regard to marine recreation. The future health of the sea is dependent upon the health of environments that are the most popular for tourism. While a number of authors are discussing the development of polar tourism as the 'final frontier' of tourism development (for example, Carvallo, 1994; Hall and Johnston, 1995), a much larger-scale tourism development with far greater implications has been occurring since the 1960s. The 'frontier', which has traditionally been the coast, is being expanded, developed and impacted upon on an ever increasing scale. Marine tourism is massive and growing quickly. From a Gaian perspective it is tremendously influential.

THE ISSUE OF ACCESS AND IMPACTS

In the past the majority of our planet's marine environment has been 'protected' from tourist use because of its inaccessibility, safety concerns and the relatively high cost of recreating in the sea. However, since the 1960s a significant number of new inventions has made the marine environment more accessible in both real and economic terms. Examples include self-contained underwater breathing apparatus (SCUBA), electronic satellite-based navigation and emergency location aids like global positioning systems (GPS) and emergency position indicator radio beacons (EPIRBs), massive aluminium high-speed catamarans, personal watercraft, windsurfers and submarines.

In addition, interest in the marine environment has grown. Television nature shows, such as those pioneered by Jacques Cousteau, magazines and films have exposed millions of people to a world that was once unknown, but is now not only interesting, but also accessible. The resultant increase in demand for marine activities has added to a tradition of sea, sand and sun holidays, and consequently millions of people now visit marine environs for recreational pursuits. There is no doubt that marine tourism is a significant issue with widespread implications for the next century:

> As the 21st century comes into focus, tourism is being revealed as a major sociocultural force with a potential to destroy, protect, or otherwise dramatically reconfigure coastal and marine ecosystems and societies.
>
> (Miller and Auyong, 1991: 75)

The causes of this increased amount and diversity of marine tourism are threefold. First, the world's population continues to grow at a rapid rate. There are now twice as many *Homo sapiens* on the planet as there were in the late 1970s – we now number around six billion. Our population will continue to increase and is predicted to reach eight billion by the year 2118 (Wright, 1990). Because the majority of this population will reside close to the coast, the use of the marine environment for recreational and other purposes will show a corresponding increase (Griffin, 1992). Second, the rapid rise in mass tourism has resulted in more people travelling away from their places of residence for recreational purposes. A significant (but as yet unquantified) proportion of this travel is to coastal areas. Of particular interest is the exposure to areas of the world that were previously undeveloped and unexposed to high levels of human use. The rise of tourism has contributed significantly to the geographical spread of human recreational activities. The third factor that has had enormous influence on the marine environment has been the invention and mass production of materials and vehicles that have improved access to and safety in marine settings. Many hundreds of machines and a wide variety of equipment now permit safe and relatively easy access to the sea. The result is an increasingly diverse range of activities in an increasing number of settings.

As the growth of marine tourism has become widespread, an increasing number of cases show that significant environmental, social, cultural and even economic damage results from tourism development and use. The question arises, therefore, of what it is that makes the difference between successful and unsuccessful tourism, particularly in the marine environment. The questions posed by Johnston reflect the concerns of many:

> Is tourism development compatible with the ideals of 'sustainable' development? Can tourism, an industry that inherently creates dependency relationships, truly be tailored in a socially responsible and environmentally viable fashion?
>
> (Johnston, 1990b: 2)

These issues, and others, are explored in this book.

THE STRUCTURE OF THE BOOK

The intention of this book is to provide a brief overview of the history and development of tourism that is based upon the marine environment, to examine the characteristics of marine tourists, and to consider the providers of marine tourism activities and opportunities. The book then looks at the impacts that these activities are having upon marine ecosystems and upon human societies. Lastly it examines the management techniques which may be appropriate to reduce the negative impacts and to maximise the benefits of marine tourism.

This work is not intended to be a comprehensive review of all marine tourism world-wide. Its purpose is to introduce the reader to a topic that has received little attention and to identify the important issues. It will, therefore, be of interest to resource managers who deal with marine environments, undergraduate students and all those who wish to gain a broader understanding of the marine tourism area, its problems, potential solutions and future challenges. It is hoped that this book will be a catalyst for further thinking and works on tourism that is based on the marine environment, for there is one thing that appears certain: such tourism will continue to grow.

REVIEW QUESTIONS

1. Debate whether you think a conservation-based paradigm is appropriate for a study of marine tourism. What are the likely advantages and disadvantages of such an approach?
2. Why is marine tourism worthy as a specialist topic for study?
3. Give an explanation of why the issue of access is so critical in studying marine tourism.

2

THE HISTORY AND DEVELOPMENT OF MARINE TOURISM

INTRODUCTION

Marine tourism has similarities with, but also differs from, the wider tourism industry. As is the case with tourism generally, tourism based upon the sea has shown a rapid increase in popularity (Miller, 1993). This rapid growth has produced significant impacts on local, regional and national communities. However, an important distinction is that marine tourism occurs (for the most part) on, in and under a medium that is alien to humans. This has a significant influence over the nature of marine tourism activities because, first, most are dependent upon equipment (such as boats and other vessels) and, second, safety issues are of greater importance. It can be argued that marine tourism is 'nature-based tourism'. For most marine tourism activities – for example, scuba diving, surfing, fishing, sailing, water skiing, sea kayaking, windsurfing – this argument is valid. However, defining some marine activities as 'nature based' is dubious. For example, on most cruise ships, where thousands of passengers holiday on floating cities, there is little direct contact with or focus on nature. Thus, when marine tourism is examined it is difficult to define tightly what it includes and its characteristics.

This chapter begins by offering a definition for marine tourism, for it is important to differentiate between it and other forms of tourism. The history and development of marine tourism are then detailed. In particular, the influence of technology is investigated, as human inventions have had a major impact on our ability to access and recreate in the marine environment. Finally, the recent rapid growth in popularity and attraction of marine wildlife is discussed.

Defining marine tourism

Whenever one attempts to define a term or topic strictly, difficulties inevitably arise. Marine tourism is no exception to this rule. There are a number of activities that are obviously within the marine tourism realm; examples include scuba diving, deep-sea fishing and yacht cruising. However, activities

Plate 2.1 Defining marine tourism is difficult. These surfers are marine tourists, but what about the cyclist on the beach?

such as fishing from a pier, exploring inter-tidal rock pools, visiting a large marine aquarium or whale watching from a headland are problematic. Humans are land-based creatures and, as a result, many of our activities that are focused on the marine environment do, in fact, occur on land. Nevertheless, the reason these activities are undertaken is because of 'marine attractions'. Thus, if marine tourism is defined too strictly, many activities which are completely dependent on things marine (such as surf-cast fishing or beachcombing) would be excluded. However, if we define marine tourism too liberally – for example, as if it only needs to be *based* upon marine attractions – a number of activities could be included which have little to do with the marine environment. Is watching a film such as *Flipper* or *Free Willy* marine tourism?

Because the majority of tourism focused upon the marine environment is based in coastal areas rather than actually on, in or under the water, to define marine tourism too tightly would be to ignore all those activities that are undoubtedly linked with marine areas. Consequently the following definition is offered:

> Marine tourism includes those recreational activities that involve travel away from one's place of residence and which have as their host or focus the marine environment (where the marine environment is defined as those waters which are saline and tide-affected).

Plate 2.2 These tourists watching seals on the rocks below are marine tourists because their focus is on the marine environment.

This definition attempts to acknowledge the term 'marine' in its biological sense – it does, therefore, exclude freshwater aquatic environments. It could be argued that sailing on North America's Great Lakes differs little as a recreational activity from sailing off the coast of New England. However, it is important to acknowledge the marine component, for marine tourism does not include exclusively freshwater-based activities like white-water rafting or stream hiking. The exclusion of freshwater environments is problematic for those marine animals which are the focus of tourism but which move between the two. For example, the West Indian manatee frequently moves from the sea up tidal estuaries and into freshwater areas. Nevertheless, the marine environment must be defined as such in order to delimit the scope of the subject of this book. It is acknowledged that this decision is largely arbitrary; however, it does use the term 'marine' in its correct sense as the basis for the definition, whilst acknowledging that not all marine tourism activities occur on, in or under the water.

This definition is explicit in restricting marine tourism to recreational activities – it does, therefore, exclude business or work-related activities. Thus commercial fishing, shipping, oil exploration and scientific research are excluded from marine tourism. It is important to acknowledge that there are difficulties in applying these distinctions absolutely; for example, there is, for many commercial fishers, a significant recreational component to their work.

The definition includes activities which are not only 'hosted' by the sea (such as windsurfing or snorkelling) but also those which have as their 'focus' the marine environment. Consequently, activities such as shore-based fishing, land-based whale watching, reef walking or watching a professional surfing competition are included, as long as they involve travel for the purpose of the activity and the focus of the activity is the marine environment or activities based on it.

EARLY MARINE RECREATION

Travel to coastal areas for the purpose of leisure has existed throughout history. Certainly many subsistence food-gathering practices, such as fishing and shellfish collection, have been major motivations for people to live in and visit marine areas. However, recreational activities such as swimming, exploring, relaxing and social events have been major attractions which have drawn humans to coastal areas for thousands of years. There are early records of activities such as sailing for pleasure:

> We know that Egyptians used sails at least in 4000 BC. Similarly, sails were rigged on Chinese junks and Viking boats in very early years. It was a development, it can be assumed, that occurred not in one place alone, but whenever man ventured on the surface of the water ... Experiencing the challenge of the wind and the water and all this meant an adventure, excitement and exhilaration; man must have acquired a love of sailing for its own sake. Boats once used solely for utilitarian purposes began to serve man for pleasure.
>
> (Brasch, 1995: 27)

There are similar ancient references to swimming and fishing for pleasure (Brasch, 1995). There is, therefore, evidence that 'marine tourism' has existed throughout history.

One particularly interesting account was contained in a letter by Pliny the Younger in AD 109 (Morris, 1988). The letter details a friendship which developed between a young boy from the small town of Hippo in present-day Tunisia, North Africa, and a dolphin named Simo. The dolphin's and boy's games and tricks became a major attraction, bringing many visitors who wished to witness the spectacle to the town. The story claims that the town became so overcrowded with tourists that major shortages of food, accommodation, water and toilet facilities resulted. Controversy over how this 'marine tourism attraction' should be managed ensued and eventually the town elders decided to take action – they killed the dolphin! This historical account illustrates that negative impacts resulting from marine tourism and debate over the management of such activities are not a recent phenomena, but have existed from the very beginning.

More detailed records regarding coastal recreation are available from the eighteenth century onwards. Prior to this time, in European societies at least, 'coastal scenery, like mountains, was not viewed with any great enthusiasm' (Towner, 1996: 169). Features that were attractive to tourists during earlier eras were those locations that featured human structures like palaces, cathedrals, cities, gardens and holiday spas. 'Wild' areas, such as the coast and sea, were viewed as barren, poverty-stricken, uncivilised places of little worth and interest (Towner, 1996). In fact bathing was regarded as immoral prior to the seventeenth century, as illustrated by the punishment issued to students at Cambridge University:

> In 1571 the Vice-Chancellor of the University of Cambridge issued a decree under which an undergraduate who was guilty of bathing in any river, pool or other water within the county of Cambridge, was to be flogged publicly in his college; and if he offended a second time, he was to be expelled from the university.
>
> (Gilbert, 1953: 12)

However, by the early 1700s an interest in and appreciation of natural scenery, including the sea, became apparent and was reflected in the art and writing of the time. By the 1800s painters such as Turner and Constable were painting coastal scenes while French poets such as Saint-Amant were praising seascapes (Corbin, 1994). An additional influence was the growth of an interest in scientific investigation of natural phenomena such as coastal geology (Towner, 1996).

The medical profession also played a significant role in prompting the growth in popularity of coastal holidays. From as early as the mid-eighteenth century some doctors advised patients to travel to the coast for their health (Gilbert, 1953). In 1750 Dr Richard Russell published his treatise on the benefits of sea water in treating diseases of the glands. He prescribed both bathing in the sea and drinking sea water as treatments for illness. Although he certainly was not the only medical practitioner to prescribe bathing in the sea for good health, Russell was the 'instigator of the seaside mania' that transformed English resorts (Gilbert, 1953: 12). Associated with an increased interest in visiting the sea for recreation during the nineteenth century was the development of transportation infrastructure, particularly rail, which made access to coastal and other natural environments far easier and more comfortable.

The growth in interest in and use of the coast during the latter part of the eighteenth century resulted in a growth of seaside holiday towns throughout coastal Europe (Towner, 1996). The fashionable elite began to desert traditional inland spa resorts like Bath in favour of seaside resorts such as Brighton. While the beginnings of sea-based tourism can be identified in England in the early part of the century, it did not occur in France until the 1780s, with Belgium, Germany and Holland following in the 1790s and Spain in the 1820s and 1830s (Walton and Smith, 1995). The United States showed a similar pattern of development to Europe, and while there are records of holiday visits to coastal locations such as Nahant and Cape May from the late eighteenth century, the development of better-known coastal holiday locations, such as Newport, Atlantic City and Palm Beach, did not occur until the mid-nineteenth century (Towner, 1996). However, whilst this growth is an important feature of the development of tourism during this time, much of the activity associated with these coastal holidays was not, in fact, based on the sea. Holidays in these locations were usually dominated by activities centred on social gatherings, entertainment – theatres and music halls, amusement arcades, parks and shopping – and for men, in some locations, public houses, gambling and prostitution (Walton, 1983). Nevertheless, the sea was a major attraction and activities such as swimming (using a variety of inventions to aid bathers), walking along the promenade and out on piers, and watching boat races were common.

Case study: Brighton, England

Brighton is one of the oldest and perhaps best-known seaside resorts in England. Although the area's growth as a resort can be traced back to the early eighteenth century, it was not until the nineteenth century that Brighton emerged as a key resort in its own right. Much of the town's development into a significant marine tourism destination can be attributed to changing aesthetic tastes, medical trends, geographic location and advances in transportation.

In 1750 Brighton was just one of seven seaside resorts in England and Wales (Towner, 1996). The town's geographic position, only 53 miles from London, gave it a strategic advantage over other resort locations (Gilbert, 1953; Towner, 1996). Its population increased by 242 per cent between 1801 and 1830, but between 1831 and 1841 the population growth slowed to only 15 per cent. This decline can in part be attributed to an 'inadequate transport capacity' (Towner, 1996: 196). However, in 1851 Brighton was by far the most populous seaside resort town, with a population of 65,569, four times larger than that of any other seaside resort town at the time (Walton, 1983).

The arrival of trains solved the transportation problem but changed the resort's mix of clientele. Prior to this Brighton's visitors were largely from the upper classes, who had large amounts of leisure time. Trains gave more working-class people, with few long holidays, the opportunity to take day excursions to the seaside. The influence of the rail link was massive. During 1837, stage coaches brought 50,000 visitors to Brighton over the whole year; in 1850 trains brought 73,000 visitors in one week.

By 1881 the residential population had grown to 107,546 and by 1911 it was 131,237 (Walton, 1983).

Brighton, unlike many other seaside resorts, developed fashionable autumn and winter seasons early in its history (Towner, 1996). This provided the tourism industry with some degree of revenue generation over the traditional off-season. As Brighton grew it developed into a 'regional service centre and residential area with commuter functions as well as being a tourist resort' (Towner, 1996: 181). This also contributed to lessening the impact of the off-season. Brighton was able as well to draw on considerable private-sector investment from outside the area, to construct such things as hotels and piers in order to keep the resort competitive (Gilbert, 1953).

By the early twentieth century there were more than 100 seaside resorts scattered throughout England and Wales, many of which were in direct competition with Brighton (Walton, 1983). Despite this competition Brighton maintained its pre-eminence in the domestic English resort market. Over the first century of its development, considerable infrastructure was developed at Brighton, so that while the resort's initial development and attractions were centred on the sea, the entertainment and social activities eventually became more important for visitors.

While the growth of seaside resort towns like Brighton have been important features in the history of marine tourism, it has been the latter part of the twentieth century that has seen the greatest change. Prior to World War II the vast majority of recreational activities associated with the sea were land based. Coastal resorts, beaches, piers and walkways hosted a limited range of marine-focused activities, such as swimming, wading, beachcombing, walking, socialising and generally relaxing, but there were few ways of actually entering the marine environment directly. Recreational boats were used in some areas, but for the most part they were expensive and relatively uncommon. It was difficult to access the sea and as a result marine tourism was constrained in its supply. However, the invention of a vast variety of vessels, machines and other technology has completely transformed marine recreation since the 1950s. The sea is no longer an inaccessible, alien environment. Humans, as a result of the influence of technology, are no longer 'land bound'.

THE INFLUENCE OF TECHNOLOGY

There are a number of inventions that are worth reviewing in detail because they have had such an enormous influence on human use of the marine world. These inventions, most of which have been developed since the 1950s, have spawned a tremendous following, resulting in millions of users and many new activities.

Scuba diving

Possibly the single most important invention with regard to marine tourism has been that of the self-contained underwater breathing apparatus (SCUBA), not only because it has allowed humans to breath underwater and has resulted in a multi-billion dollar industry, but more importantly because it has facilitated a changing of attitudes about the marine world. Prior to the advent of scuba the undersea world was largely mysterious and unexplored. This uncertainty and inaccessibility had a widespread impact on the willingness of people to participate in activities on the water. The impact of photographs, books and especially television shows like *The Undersea World of Jacques Cousteau* has been massive. This exposure of the marine environment and the things that live within it to a wide audience would not have been possible without scuba. An increased understanding of and interest in things marine generated a demand not only for scuba diving, but for general exploration and enjoyment of the sea. This demand has, in turn, resulted in the invention of many more ways of accessing the sea for recreation. Scuba, therefore, changed the image of the sea from an alien, inhospitable and threatening location to a fascinating, enjoyable and, most importantly, accessible one. (For more information see Chapter 3, 'Demand'.)

Mixed gas diving and rebreathers

The most popular way of visiting the underwater world is by using scuba or snorkelling. However, traditional scuba diving and even more recent technology such as mixed-gas-based aqualungs (e.g. those using nitrox) have been constrained by depth and time. Decompression sickness – 'the bends' – and nitrogen narcosis restrict the time divers can spend at depth; consequently most recreational diving occurs less than 30 m below the surface. The recent invention of computer-controlled, closed circuit underwater breathing systems known as 'rebreathers' has the potential to revolutionise the use of the underwater environment. The rebreathers recycle the gas being breathed by the diver, using chemical scrubbers that cleanse the gas of carbon dioxide. A computer controls the mix of gases delivered to the diver in order to avoid problems posed by depth and pressure:

> With this technology, dives of up to 120 metres can be safely made, even in remote locations, without the need for prohibitively expensive back-up facilities at the surface. Divers can also extend their stay at shallow stops with a degree of safety previously unavailable.

> An added advantage is that rebreathers shed no bubbles except during ascent.
>
> (Anonymous, 1991: 10)

Rebreathers will provide a much greater level of freedom for divers and allow the safe exploration of a much greater area underwater than was previously available. In addition, they are quieter (no bubbles) and consequently divers will find it easier to approach aquatic animals.

Accommodation

Underwater accommodation is now being offered to tourists visiting Key Largo in the Florida Keys (Jones, 1993), and a 'Flotel' (floating hotel) which was tried on the Great Barrier Reef (Kelleher, 1990) is now based in a harbour in Vietnam. Water-based accommodation on a growing number of boats, including a multitude of small pleasure craft as well as many large cruise ships, means that tourists are increasingly able to stay overnight on (and even under) the water.

Passenger-carrying vessels

Changes in technology have also impacted on larger passenger-carrying vessels, in particular in terms of their speed. Many offshore attractions such as reefs and islands were, in the past, difficult to access because they required a long, often uncomfortable trip. In the early 1980s the introduction of high-speed catamarans capable of cruising at 25–30 knots brought many offshore locations within reach for day-trip-based tourists. The impact of such vessels on attractions like Australia's Great Barrier Reef has been massive:

> Until 1982 only Green Island and two or three other reefs were within 60 to 90 minutes travelling time of Cairns ... For much of the year that journey was uncomfortable because of prevailing weather conditions so the Great Barrier Reef was virtually inaccessible to the majority of tourists. The introduction of comfortable high speed catamarans brought some 46 reefs within day trip reach of Cairns or its northern offshoot, Port Douglas, and generated a major increase in reef visitor numbers.
>
> (Kenchington, 1990a: 27)

Boat builders are currently working on 'surface-tension'-based superfast boats that will be capable of speeds of 45 knots. The introduction of these kinds of vessel will, once again, significantly increase the ability of tourists to access offshore and remote locations. Amphibious land, water and submersible craft are also in operation and it is likely that additional advances in these technologies will provide far easier access into, across and under the water.

Recreational vessels

Another important influence has been the advent of mass-produced, relatively cheap, reliable boats. Prior to the 1950s, boats were built mainly out of wood by experienced craftspeople. A significant amount of time and money was involved in the production of a vessel, and consequently, those boats that were used for recreational purposes were predominantly owned by the wealthy. The invention and use of fibreglass, aluminium, ferro-cement, inflatable hulls and polyethylene in recreational vessels has transformed boat ownership. Not only are these boats able to be mass produced and bought relatively cheaply, they have proved more reliable and more seaworthy, have greater longevity and require less maintenance than their earlier wooden counterparts. Wooden boats have also remained popular because new materials (for example, plywood) and construction techniques have produced similar advantages. Boat ownership has grown massively since the late 1960s, and as a result the sea has become more accessible to a greater number of people.

Boats form the basis for a wide variety of other marine recreational activities, including fishing, diving, water-skiing, sightseeing, racing and whale watching. Boats also provide a means of transport to previously inaccessible islands, reefs, harbours and beaches. Different types of vessel have added to the options available. Some, such as sea-kayaks, air-boats, jet-boats, personal watercraft (such as jet-skis) and hovercraft have allowed access to areas which were previously difficult to get to. Others, such as sailboards and sail boats, have spawned entire new recreational and competitive sports. A significant change in the use of powered boats was the invention of the outboard motor. Lighter, more powerful and cheaper to run, the outboard motor has allowed an entirely different approach to boat design. Boats can now be small, light and open decked, in contrast to the larger, heavier type of vessel needed to house an inboard engine. Thus, changes in boat-building technology, design and power have had a significant influence on the rapid development of marine tourism.

Surfboards and sailboards

Surfing has had a massive influence on the image of marine activities, and forms a world-wide recreational activity participated in by millions. The exact origins of surfing are not clear. It is likely that it was practised by many coastal cultures surrounding the Pacific Ocean for many centuries. Accounts from Captain Cook's visit to the Hawaiian Islands in 1778 confirm that it was a widespread practice then, and it is thought to have been practised as far back as the tenth century (Brasch, 1995). Modern fibreglass boards, films, music, clothing and a professional circuit for top competitors have helped establish a 'surfing culture' which has spread world-wide since the late 1960s. There are now millions of active surfers. A number of coastal locations, such as Oahu's North Shore in Hawaii, Tuvalu in Fiji and Jeffereys Bay in South Africa, have developed a significant tourism industry based upon their attraction as surfing locations.

In a similar way to surfing, sailboarding or windsurfing has attracted a growing clientele. The first windsurfer was invented in the 1970s and since then the sport has expanded and spread to virtually every coastal area. In a pattern that mimics the popularity of surfing, a sailboarding 'culture' has developed and many locations such as Hookipai on Maui, Hawaii, have become sailboarding tourism destinations.

The submarine

An invention originally conceived for military purposes but now also used for recreation is the submarine. Although the use of personal submarines is not yet widespread, a number of locations feature operators who offer underwater trips to tourists in submersible and semi-submersible craft (see case study below). It is likely that the use of this invention will increase and offer yet another means of accessing the underwater environment.

Navigational aids

A further significant recent advancement has been the production of increasingly accurate navigation, communication and safety equipment for use at sea. Computers for scuba divers which calculate depth, water temperature, air remaining, decompression times and navigation information have been widely available since the late 1980s. Similar changes in electronics for boats have rendered navigation far easier and safety far more attainable than ever before. GPS, which now only cost a few hundred dollars, are able to locate the user to within several metres almost anywhere on the planet. Similarly, EPIRBs are readily available and facilitate the quick location and rescue of vessels in distress. Depth sounders, fish finders, electronic chart plotters and logs, weather faxes, weather satellite images and many other devices are now commonly used by people on boats.

The influence of these inventions has been massive. The proliferation of new equipment which is mass produced and marketed has continued to increase since the revolution provided by the invention of scuba in the late 1950s. It is almost certain that this trend towards large numbers of mechanisms for accessing the marine environment will continue to increase. Consequently, more people will be using the marine environment for a greater variety of activities in the future. Many marine-tourism-based companies have been founded on these new technologies. One such company, which illustrates the commercial potential of marine technology, is Atlantis Submarines.

Case study: Atlantis Submarines

Atlantis Submarines Incorporated was the first company to apply submarine technology commercially to the tourism field. A submersible for passengers was tried in Hawaii in the early 1980s and proved so popular that the company now operates over 15 submarines and several semi-submersibles in Aruba, Grand Cayman Island, the British Virgin Islands, the Bahamas, St Thomas, the US Virgin Islands, Cancun (Mexico), Guam and Barbados as well as in Hawaii.

The success of Atlantis is impressive and is indicative of the influence of new technology applied to an opportunity for marine-based tourism. In a little over its first 10 years Atlantis took over 4.5 million tourists on over 200,000 dives. This produces an annual turnover for the company of around $US75 million. The company now employs over 500 people and caters to almost a million tourists each year (O'Halloran, 1996).

The submersibles and semi-submersibles operated by Atlantis have proved popular because they have made access to the underwater environment easy, comfortable and dry. Prior to the use of submersibles the undersea environment was only accessible via scuba diving or snorkelling, which appeal to a limited market; in glass-bottom boats, which have limited viewing opportunities as they are restricted to from-the-surface observations; and in a few underwater observatories, which do not offer mobility. The submersibles have proved reliable and are able to be used frequently, and the company adds in a dive package as part of the entertainment for passengers. The following account illustrates the type of trip offered:

> Twelve dives are undertaken daily and as part of the services offered there is the unique premium Atlantis Submarine tour which begins with a close up view of the most scenic areas around Fresh and Bay Reef and climaxes with a unique and exhilarating dive show. The live 15 minute interactive dive show is both dramatic and educational. Atlantis divers, Flash and Ray, with the use of advanced sub-aquatic technology, communicate directly with Atlantis and her passengers as they glide with underwater scooters creating a thrilling underwater choreography.
>
> (*Caribbean Weekly – Barbados*, October 1995)

The success of Atlantis is attributed to careful selection of locations for operating (including both operational and political considerations) and to a close analysis of market trends in each location and of tourism cycles globally. In addition, Atlantis has recognised the need to diversify and offer services additional to their core business of submarine tours. Boat tours, scuba and snorkelling trips, and sailing cruises are now also offered in a number of locations. Co-operative marketing arrangements have also been fostered with coastal resorts and hotels as well as cruise ship companies (O'Halloran, 1996).

Atlantis has, therefore, utilised new technology to establish and develop a successful marine tourism business that is now a major provider of employment, revenue and opportunities for tourists in Hawaii and the Caribbean.

Marine tourism has, therefore, expanded dramatically over the past decade. This diverse range of activities occurs in a wide variety of locations for an equally wide variety of reasons. Technological changes have allowed humans access to areas of ocean that were previously little used. Within easily accessible areas many more activities are undertaken.

THE DIVERSITY OF MARINE RECREATIONAL ACTIVITIES

When the development of marine tourism is traced, the most obvious features that emerge are, first, that it has become increasingly popular and, second, related to that popularity, that it has become increasingly diverse. Today there are more ways and means of accessing the marine world for recreation than ever before. It appears highly likely that there will continue to be an increase in this diversity as we invent more ways to go under it, get in it and get on it. This complex array of activities, some complementary and many not, create many management challenges as agencies responsible for marine resources try to reduce conflicts, decrease risks for participants and minimise damage to natural resources.

Within this trend of increasing diversity and increasing popularity several important patterns can be identified. First, as one would expect, greater use tends to occur close to areas of human concentration, namely cities. Despite the heavy emphasis on analysing long-distance travel in tourism research and literature, the majority of recreation occurs close to people's place of residence, and this is also true of marine recreation. So whilst much of this book concentrates on the more commercial aspects of marine tourism involving longer-distance travel away from home, it must always be remem-

Plate 2.3 Even remote areas are now accessible as a result of changes in technology, particularly in transport.

bered that the great majority of marine recreation is a result of regular 'day-trip' recreation from local residents. This point is emphasised in an observation by Miller:

> It should be kept in mind that six out of ten people around the world reside within 60 kilometres of the coastline and two-thirds of the world's cities with populations greater than 2.5 million are located by tidal estuaries. The population of the coastal zone is projected to double within the next 20–30 years.
>
> (Miller, 1990: 6)

Because much marine recreation occurs in close proximity to urban areas, the environments upon which the recreation is based are subject to increased pressure. Marine environments closer to cities receive large amounts of urban 'run-off' and other discharges resulting from human activity (sewage, stormwater, etc.). In addition, these environments are often subject to dredging, foreshore alteration and reclamation, and dumping of waste products. They are more frequently fished, and have larger numbers of vessels and navigation aids. The effect of this higher level of use is that these areas are more vulnerable to additional pressure such as that produced by recreational activities.

A second main trend that is easily identified with regard to the spatial distribution of marine recreation is that there is an inverse relationship between distance from shore and intensity of use. A wider variety of activities and a greater level of use are associated with near-shore environments. Again, this pattern is entirely logical, for humans are terrestrial-based animals who find it difficult to survive in a liquid medium. It is important to recognise the dilemma this causes. Greater levels of human use occur close to shore and close to cities, and it is these environments that are the most critical to the health and long-term survival of the oceans and all that live within them (Earle, 1995). Marine tourism predominantly occurs, therefore, in those ecosystems that are most vulnerable to disturbance.

The sum of these aforementioned trends, in crude terms, is a massive increase in use. Marine tourism has exploded in the past few decades and is now a significant social and economic force world-wide.

THE GROWTH AND IMPORTANCE OF MARINE TOURISM

The widespread and rapid growth of tourism generally is well documented (see above). It is not surprising therefore that marine tourism shares this trend and is also predicted to increase rapidly (Miller and Kaae, 1993). It is difficult to separate marine tourism from general tourism data, for as Miller (1990: 1) notes: 'Unfortunately, government and industry travel statistics are generally not compiled in a manner which clearly documents the nature of coastal zone tourism.' There appears to be, however, a consensus in the literature that coastal and marine tourism is growing at an even faster rate than the general tourism sector (Miller, 1990). This growth reflects not only increasing opportunities for marine recreation but also a 'generally increased level of interest in anything to do with marine environments' (Shackley, 1996: 111). A limited number of studies suggest that the growth of marine tourism has been relatively recent. For example, research on marine tourism businesses in New Zealand revealed that over 60 per cent of operators (400) had been in business less than five years (McKegg, Probert, Baird and Bell, 1996). Certainly for many island nations marine tourism is the mainstay of the local economies. For example, the Seychelles, a small island nation in the northern Indian Ocean, derives approximately 70 per cent of its foreign exchange earnings from tourism (Gabbay, 1986), and this tourism is 'entirely ocean based' (Sathiendrakumar and Tisdell, 1990: 79). In Bermuda, approximately 40 per cent of public revenue is derived from tourism and tourism businesses generate over $US1,225 million (Archer, 1989).

An important indicator of the relative importance of the sea as a tourism attraction is shown by a recent study on the value of beaches in the United States. Houston found that:

> Beaches are key to U.S. tourism, since they are the leading tourist destination, with historical sites and parks being second most popular, and other destination choices minor by comparison. Coastal states receive about 85 per cent of U.S. tourist-related revenues, largely because of the tremendous popularity of beaches. For example, a single beach, Miami Beach, has more annual visits than Yellowstone, the Grand Canyon, and Yosemite National Parks combined.
>
> (Houston, 1996: 24)

Table 2.1 Forecast for organised watersports-based tours worldwide 1994–2000

	1994	*1996*	*1998*	*2000*
Estimate of total number of watersports-based tours	0.75–1 million	1–1.5 million	1.5–2.5 million	2.5–5 million
Relative market share of different activities (%)				
	1994	*1996*	*1998*	*2000*
Bareboat	40	33	30	25
Flotilla	9	10	10	10
Crewed	3	4	5	5
Diving	14	20	25	30
Shore based	34	33	30	30

Source: Smith and Jenner, 1994

This finding is even more significant when one considers that the United States is by far the world's most important tourist destination (Waters, 1990).

Smith and Jenner (1994) estimated that watersports-based 'package tours' would increase by between two and a half and five times by the year 2000. Table 2.1 gives a breakdown of the market share of differing types of water-based packaged tours and reveals the growing predominance of shore-based and diving-focused marine tours.

Many nations with significant coastlines are less developed, have small or insignificant tourism industries and have little infrastructure to support tourism. However, many of these nations are viewing tourism as a catalyst for economic and social development (Pattullo, 1996). In particular, a number of island nations hope to develop tourism based upon their many natural areas which are undeveloped, unique and 'unspoilt', and hence of immense value and interest (Lockhart and Drakakis-Smith, 1997). Additionally, marine flora and especially fauna are a major draw for visitors interested in nature-based attractions. Much of the tourism growth predicted for island nations is, therefore, likely to be associated with natural environmental features. The potential for marine tourism is enormous globally.

However, while the general trend for tourism is one of increasing numbers and increasing influence in coastal and island nations, a number of examples reveal the fickle nature of the tourism industry. Political unrest, such as the 1987 coup in Fiji, can have a significant influence on tourism – it fell there by 26 per cent (Waters, 1990). Climatic events, such as Hurricane Hugo in the Caribbean in 1989, or the loss of a significant carrier, such as the bankruptcy of Eastern Airlines in the same year, can have a significant negative influence on tourism arrivals, as it did in the Caribbean in 1989 (Miller, 1990). Despite the size and growth of marine tourism, the industry still tends to be more weather influenced than general land-based activities.

Case study: the cruise-ship industry

During the 1920s cruises were the preferred mode of travel for the world's social elite. Sea travel at this time remained the only practical means of travelling over large expanses of water. Following World War I America imposed restrictions on the flow of immigrants permitted to enter the country. Many of the ships that had been used to transport immigrants were now redundant and as a result were refitted into cruise ships (Lundberg and Lundberg, 1993). It was not until the arrival of the passenger aircraft that cruise ships were rendered obsolete as the primary mode of travel.

In the early 1980s cruise ships again came into vogue, aided by aggressive marketing campaigns and TV shows such as *The Love Boat*. For the first time cruises were pitched at the middle-income population, not only the very rich. This was achieved by shortening cruises, introducing fly-cruise options and increasing ship capacities. As a result cruise passengers have also started getting younger: approximately half are under 45 years of age, with one third being under 35 (Lundberg and Lundberg, 1993).

Since cruising's renaissance in the early eighties the industry has been one of tourism's leading success stories. It is dominated by three cruise lines: Carnival, Royal Caribbean and P&O. It is predicted that in the future these three companies will increase their dominance of the industry via new ship orders rather than through mergers with other cruise lines (Peisley, 1995). Advance ship orders indicate that cruise lines are investing in increasingly larger ships, some of which are 100,000 tonnes (Peisley, 1995). In 1998 at least 28 new ships were due to be delivered to cruise lines (Pattullo, 1996).

The most popular cruise destination in the world is the Caribbean. This popularity can be attributed to the region's ideal cruising environment and its close proximity to the world's largest cruise market, the United States. The port of Miami serves as the major gateway to the Caribbean and is the largest cruise port in the world. In 1993 the Caribbean routes were plied by 77 cruise ships with passenger capacities averaging more than 1,000 each (Pattullo, 1996). The region's most popular port of call, the Bahamas, recorded 1.8 million cruise-ship arrivals in 1994 (Pattullo, 1996). It is predicted that the Caribbean will maintain its dominance as a cruise destination as many companies still see an untapped potential in the area. Other significant cruising areas include the

Mediterranean, the Pacific, Alaska, North Africa and the South China Sea. Cruise lines often operate their ships in different regions depending on season.

By the turn of the century it is predicted that the cruise industry will carry in excess of 8 million passengers annually (Peisley, 1995). This would make cruising one of the world's leading tourism industries.

The case of the cruise-ship industry provides further evidence to support the conclusion that marine tourism is increasing rapidly. (For more information see Chapter 3, 'Demand'.) An additional segment of marine tourism which has also shown spectacular growth in the past decade is that focusing on wildlife.

WILDLIFE AND MARINE TOURISM

People have always been interested in animals, as the fact that domestic pets have been companions of humans throughout history and across many cultures illustrates. Actually, knowledge of animals and their behaviour has probably been a central part of human life, because animals have provided much of the food supply for most human societies. Today, many indigenous peoples continue to interact with wildlife for spiritual and cultural reasons as well as for food. However, the idea of visiting and observing wild animals for recreational purposes, as a tourist attraction, has been a more recent phenomenon (Shackley, 1996). Zoological gardens began to appear in the nineteenth century as European explorers began to bring specimens back from their travels (for example, Anderson, 1878). Safaris to view and hunt wildlife, in locations such as Africa, began around the same time (Adler, 1989).

Since the late nineteenth century, the growth of facilities which hold wildlife captive and the management of locations which protect wildlife have been widespread (Yale, 1991). Many large cities throughout the world now have zoos – the 1982 *International Zoo Yearbook* listed 757 zoos worldwide (Yale, 1991) – and many countries manage national park networks which protect wildlife and facilitate their observation. This range of opportunities for tourists to interact with wildlife has continued to increase, and consideration of how these interactions should best be managed has become a common topic within the tourism and wildlife management fields (for example, Vickerman, 1988; Shackley, 1990, 1996; Kerr, 1991; Albert and Bowyer, 1991; Duffus and Dearden, 1993; Orams, 1995).

Vickerman (1988) proposes that wild fauna are particularly important tourism attractions, and illustrates this by stating that some $US14 billion is spent annually on wildlife viewing, photography, travel and feeding. Rockel and Kealy (1991) report on an earlier study in 1980 that found around 29 million people took trips specifically to interact with wildlife in the United States. Wildlife interaction (observing, feeding, touching, photographing or otherwise experiencing wild animals) occurs in a wide variety of settings and, in recent years, has become increasingly popular (Duffus and Dearden, 1990; Clamen and Rossier, 1991; Duffus and Wipond, 1992; Heath, 1992; Muir, 1993; Hammit, Dulin and Wells, 1993). However, we have a poor understanding of the role of these various interaction opportunities in outdoor recreation experiences (Shaw, 1984).

There is a similar pattern of interest in marine fauna. McKegg and colleagues (1996), in a study of marine tourism businesses in New Zealand, found that animals, particularly sea birds and marine mammals, were the prime attraction for over 65 per cent of businesses. Similar patterns of interest in marine wildlife are noted for other locations; for example, Forestell and Kaufman report on the popularity of whale watching in Hawaii: 'A survey of major operators indicates that an estimated 130,000 people actually went whalewatching during the 1990 season, with approximately 110,000 going whalewatching from Maui alone' (Forestell and Kaufman, 1990: 401). Another notable example is the dolphins of Monkey Mia in Western Australia:

> The wild bottlenose dolphins have been visiting the beach since at least 1964 with up to twenty arriving at a time up to six times a day for handouts of fish. This phenomenon is attracting lots of tourists with over 100,000 visitors in 1990.
>
> (Dowling, 1992: 131)

Plate 2.4 Marine mammals and other charismatic wildlife are powerful images used to attract marine tourists.

Interacting with marine animals in natural or 'wild' settings is, therefore, a major tourism attraction. These interactions typically involve greater effort by the tourist, or, in most situations, by the tourist operator, to view the animal in its natural setting as opposed to a captive (oceanarium) situation. The most common locations for these interaction opportunities usually centre on breeding sites such as the Mon Repos turtle-breeding beaches on the central Queensland coast in Australia (Cherfus, 1984), along migratory routes like those used for whale watching in southern Queensland, Australia (Jeffery, 1993) or at feeding sites such as that provided by the continental-shelf-induced upwelling off the coast of Kaikoura, New Zealand (Baxter, 1993).

There are also a number of situations where animals are manipulated in some way in order to allow tourists an opportunity for regular or closer interaction. This is usually done through the feeding of the wildlife at a fixed location at a regular time. Australian examples include the feeding of dolphins at Monkey Mia in Western Australia (Connor and Smolker, 1985) and at Tangalooma on Moreton Island in Queensland (Orams, 1994), and the feeding of potato cod at Cod Hole on the Great Barrier Reef (Adler, pers. comm.).

There is no doubt that interacting with wildlife, in the wild, has become extremely popular. The rapid growth of the whale-watching industry is a case that aptly illustrates the potential of tourism based upon marine wildlife in its natural environment.

Case study: whale watching

The transition from an industry which was based upon killing cetaceans (whales, dolphins and porpoises) to one based on observation and interaction with them is symbolic. For many coastal locations, whale-watching tourism has proved more economically sustainable than commercial whaling (Orams and Forestell, 1995). The rapid growth of the whale-watching industry illustrates the huge attraction provided by large, charismatic wildlife – alive and in its natural state.

In the early 1980s there were only around a dozen countries which conducted commercial whale-watching activities. These whale watches were predominantly based upon taking tourists out to sea in big vessels to observe larger baleen whales (such as humpbacks, grey and right whales) when they came to the surface to breathe. Today 295 communities in over 65 countries host whale watching. The growth of the industry has been spectacular. Data available from the early 1990s show it has been increasing at around 10 per cent per year. In 1994, an estimated 5.4 million tourists went whale watching, generating over $US500 million in revenue (Hoyt, 1996). Watching whales (and dolphins) is, quite simply, big business. In many cases it has transformed small, economically depressed coastal towns into thriving tourism centres (see case study on Kaikoura, Chapter 5).

As the industry has developed, the number of species targeted has expanded. Additional cetacean species such as those in the toothed cetacean group are now the basis of commercial tourism operations. Examples include sperm whales, orca or killer whales, and bottlenose, dusky, spinner, common and Hector's dolphins. Operators have also started to explore additional possibilities for interaction, no longer confined to boat-based observation. In some situations shore-based feeding stations have been established. Swimming with wild dolphins using snorkel and masks, underwater motorised 'scooters' and tow lines is now popular. Observation now occurs from a greater variety of vessels including cruise ships, sail boats and sea-kayaks.

As a result of long-term experience with whale watching in Hawaii (one of the first locations to begin commercial whale watching), Forestell and Kaufman (1995) suggest that its development follows a four-stage cycle of discovery, competition, confrontation and stabilisation. This model is a particularly interesting one to test in other locations where controversy exists over the management of the industry.

The development of the whale-watching industry illustrates the pattern of many sectors of the marine tourism industry. Initially the industry developed as a result of a recognition of an opportunity provided by local conditions (in this case, the presence of large whales in a relatively protected coastal area). The success of early operators attracted additional investment and expansion as other suitable locations were identified, and later a significant diversification of the 'product' occurred as new ideas and technologies were applied.

In addition to the observation of 'wild' marine animals there are a number of facilities which are 'captive'. Some have large, fenced-off areas of water within which marine mammals and other marine wildlife can be viewed. In some cases, tourists are permitted to swim with them (Doak, 1984). There are also those facilities which are involved in research and/or rehabilitation of injured, orphaned or sick wildlife and which are open to the public. A number of these types of operation exist for birds, marine mammals and other wildlife (Yale, 1991).

There is, therefore, a wide range of opportunities for tourists to interact with marine animals, and this aspect of marine tourism is proving extremely

Plate 2.5 Watching whales has become big business world-wide.

popular. In particular, those animals that are unusual and/or endangered are especially targeted (Shackley, 1996). Concerns are being expressed about the potential impact of this increasing industry on the animals. For example, Forestell and Kaufman observe that controversy has resulted from the rapid growth of humpback whale watching in Hawaiian waters: 'Concern has grown in every quarter that the cumulative effect of this activity may threaten the recovery and survival of this endangered species' (Forestell and Kaufman, 1990: 401). A number of authors raise ethical issues regarding the justification of wildlife-based tourism. For example, Guthier states: 'While there are many advantages for humans in the recreational enjoyment of wildlife, I am hard pressed to conceive of any advantages for the wildlife' (Guthier, 1993: 98). Consequently, there is a need to establish what impacts result from marine-wildlife-based tourism (Chapter 5) and a further need to develop management strategies which mitigate those impacts that are negative and expand those which are positive.

SUMMARY

This chapter has offered a definition of marine tourism, briefly reviewed its history and traced its recent development. While there are few data available on sea-based tourism activities, there are a number of studies which suggest that it is growing rapidly, at an even faster rate than the general tourism industry. The case studies on cruise ships and whale watching provide additional evidence which confirms that the development of marine tourism has been rapid. One of the major influences producing this growth has been the invention of new mechanisms for accessing the marine world. Technological changes have opened up a range of opportunities for marine recreation that did not exist 30 years ago. In addition to this expansion on the supply side (investigated further in Chapter 4) there has also been a rise in interest in things marine. The implications of this rapidly increasing use of the sea for recreational purposes are important because much of this tourism occurs in areas already heavily stressed by human actions. Furthermore, many marine tourist operations are based in areas or on animals that are endangered and/or sensitive to disturbance. Consequently a strong case can be made for the need to investigate marine tourists and marine tourism operators further, particularly with regard to assessing impacts and the success of various management strategies. The remainder of this book attempts to do this, first by outlining what we know about marine tourists (the next chapter) and then looking at the range of marine tourism opportunities available (supply) and finally at the impacts and management of the industry.

REVIEW QUESTIONS

1. Discuss why defining marine tourism is difficult.
2. Explain the influences that changed attitudes about the sea in seventeenth- and eighteenth-century Europe.
3. Name three important inventions that have had a significant influence on the development of marine tourism. Explain their impacts.
4. Why do you think marine wildlife has become such an important attraction for tourists?
5. What do you think will happen to marine tourism in the future?

3

WHO ARE MARINE TOURISTS?

INTRODUCTION

In the previous chapter an introduction to the size and scope of the marine tourism industry was outlined. In this chapter, the characteristics and qualities of marine tourists themselves are explored. Understandably, because the types of recreational activity undertaken in the marine environment are diverse, so are the participants in those activities. However, there is evidence which shows marine tourists do differ demographically from the general population, particularly when specific activities are considered. It can also be argued that the motivations of marine tourists may be different from those that are detailed for the wider tourism industry. These motivations, and the experiences that participants seek, are directly influenced by the medium in which the activities occur, namely, the sea.

MOTIVATIONS

It is important to understand why individuals undertake certain activities, for if we are able to understand that, we are far better able to plan for their needs and to manage their actions. 'Psychologists agree that a motive is an internal factor that arouses and directs human behaviour' (Iso-Ahola, 1989: 248). Discovering people's motivations is, of course, extremely difficult. Motivations cannot be observed, they can only be inferred from observing behaviour. To discover motivation, subjects must be asked for their reasons for actions, and these are usually complex and are seldom fully understood by the actors themselves, so even asking people about their motivations is problematic. Despite this, the concept of motivation for recreation or leisure has been widely explored in the literature (Ewert, 1989) and a wide variety of theory has been developed. There is not, however, one theory or a set of theories which adequately explains why humans do what they do. Nevertheless, a number of these theories are worthwhile reviewing as they have relevance to marine recreation and tourism.

A number of authors have argued that, at a basic level, humans are motivated to travel by curiosity (Crompton, 1979). The desire to see what is 'over the horizon' (or perhaps in the marine situation what is 'under the horizon') may be an important reason why individuals undertake exploration-based activities such as sea-kayaking, scuba diving, snorkelling and yacht cruising. This curiosity motivation was incorporated by Gray (1970) in his 'wanderlust'–'sunlust' theory. Gray contends that the motivation for leisure travel can be explained by a desire to escape the familiar and experience different environments, activities and people. Related to this is 'sunlust' – the seeking of a setting that is perceived to be better for particular recreational activities. These two concepts have obvious application to the marine tourism situation because, first, marine environments are different and offer an opportunity to escape normal routines and surroundings. Second, marine recreational activities are setting dependent (that is, you cannot go water-skiing in your backyard) and heavily influenced by weather conditions. Thus, the motivation to travel for surfing enthusiasts is to go to a marine location where they can undertake the activity they have a passion for.

A related theory is that proposed by Iso-Ahola (1982), who argues that motivation for leisure can be explained by two fundamental forces: seeking personal and social rewards and escaping from personal and social environments. Intrinsic rewards are sought in leisure in the sense that participants derive satisfaction from such things as the challenge or excitement of the activity, as in, for example, hooking and landing a large bill-fish. What may also be important is that the activity affords an opportunity to escape from the regular routines of life, such as the rigours of work. An additional component recognised in the model is that social factors are also important motivational influences. It is well recognised that most leisure experiences have a social component and that these interpersonal aspects are important influences on participants' enjoyment (Crandall, Nolan and Morgan, 1980). What is not often realised is that many leisure activities are undertaken because they offer the opportunity for participants to escape from social settings or obligations. The age-old justification for holidays to 'get away from it all' therefore includes not only getting away from activities and settings, like home and work, but also getting away from people, like workmates, families and other relations.

The application of Iso-Ahola's approach to marine-based leisure activities is valid. Activities such as fishing, diving, sailing, surfing and so on are attractive not only because they are inherently enjoyable for participants but also because they offer an opportunity to 'escape' from other environments, both physical and social.

Much of the early work on motivation was based on an extension of humans' need for physiological balance or 'homeostasis'. It was proposed that a person's actions are motivated by a need to keep a balanced level of psychological arousal or stimulation, in the same way as the human body is driven to maintain a constant body temperature or energy level (Duffy, 1957). This idea was supported by studies which showed that both understimulation and overstimulation were psychologically harmful to humans (Hunt, 1969). Thus, people are motivated to attain stimulation and excitement in their leisure activities when they are bored or understimulated in their other endeavours (work, family life, etc.), whereas those who are overworked, stressed and overstimulated are motivated to find relaxation and passivity in their leisure.

Certainly the variety of marine recreational activities shows varying levels of relaxation and excitement. Many, in fact, offer both. Fishing, for example, typically involves long periods of inactivity, affording the opportunity for relaxation, while also offering periods of excitement when a fish is hooked and is being landed. While the 'optimum level of arousal' idea acknowledges that 'the optimum level varies from person to person and from time to time' (Iso-Ahola, 1989: 249), it is inconsistent with the fact that there are a number of people who appear to need stimulation and action almost constantly, whilst others appear to need little or no activity at all.

A related idea that has relevance to a number of marine recreational activities is Csikszentmihalyi's (1975) 'optimal experience theory'. Csikszentmihalyi contends that when participants engage in an activity which matches their skill level with the challenge provided by the activity, a psychological state, which he terms being 'in the flow', is achieved. This state is characterised by the participant's concentration on the task, a loss of a sense of time, feelings of competency and euphoria. In contrast to the positive psychological state of being 'in the flow', if the skill level of the participant exceeds the challenge level of the activity, boredom results. Conversely if the level of challenge provided by the task is higher than the skill level of the participant, stress and anxiety result. This 'flow' state has also been termed 'peak experience' and studied in risk recreation (Haddock, 1993). Closely related is the work of Zuckerman (1971, 1979, 1985) on 'sensation seeking' and the early work of White (1959) and Deci (1975), who argued that there is a basic human need for feeling empowered and capable. The result, according to these authors, is that humans seek out situations where they can be challenged and derive a sense of achievement in their mastery of the task or environment.

These ideas of 'peak experience' or being 'in the flow' and having a need for feelings of competence are of interest in the marine situation, because most marine recreational activities contain an element of challenge and risk and it is obvious that this is part of the attraction for participants. The work of Csikszentmihalyi, Iso-Ahola and others helps to explain this desire. Marine settings and activities provide tourists with an opportunity to be challenged and to experience this 'flow' state that appears to be so beneficial. In addition, marine activities allow participants to accomplish things that result in feelings of achievement and competency. Furthermore, marine settings provide an environment that is different, a situation which many tourists seek. It also provides an opportunity to escape from daily routines and commitments. Many of these motivational theories can be recognised as influential factors in the following quotation about fishing:

> The objectives of recreational fishing are complex but it is clear from several studies that the experience of attempting to catch fish, the pleasures of boating, and escape from work and domestic routine are generally more important than the reality of catching large amounts of fish. The element of challenge of personal skill and determination against a cunning adversary in its natural element is important, and is reflected by the scorn with which anglers greet fish that don't fight.
>
> (Kenchington, 1990b: 26)

An additional relevant issue has been the recognition of humans' needs for solitude, peace and closeness to nature (Frigden and Hinkelman, 1977; Hendee and Roggenbuck, 1984). Wilderness experiences have become increasingly sought-after features of recreational activities in developed countries like the United States (Miles, 1990). Early writers on the value of nature-based wilderness experiences have influenced a conservation movement which has grown since the nineteenth century. In 1853 Henry David Thoreau expressed a need with which many millions have identified, when he stated:

> I long for wildness, a nature which I cannot put my foot through ... This stillness, solitude, wildness of nature is a kind of thoroughwort, or boneset, to my intellect. This is what I go out to seek.
>
> (Thoreau, 1972: 18)

Thus, for many, the opportunity to recreate in a natural setting, uninfluenced by human activity, is extremely important. Scherl (1987) argues that there are significant psychological benefits that arise from wilderness experiences. There is no doubt that, for many, wilderness experiences are provided by marine-based recreation, for it is the sea that provides one of the few truly wild areas left on our planet. However, the opportunity for peace, solitude and 'wildness' is decreasing, even on the sea. Aldo Leopold's comments of half a century ago are even more relevant today:

> Wilderness is a resource which can shrink but not grow. Invasions can be arrested or modified in a manner to keep an area usable for recreation, or for science, or for wildlife, but the creation of new wilderness in the full sense of the word is impossible ... One of the fastest-shrinking categories of wilderness is coastlines. Cottages and tourist roads have all but annihilated wild coasts ... No single kind

> of wilderness is more intimately interwoven with history, and none nearer the point of complete disappearance.
>
> (Leopold, 1949: 194)

The issue for management is clear: wilderness experiences are desired by a segment of the marine tourism market, but opportunities for such experiences are decreasing. Resolving such issues is discussed in more detail in Chapter 6.

In summary, it is fair to conclude that humans are complex animals. Their motives for what they choose to do in the marine environment are equally complex. However, a number of theories about motivation, including wanderlust–sunlust, seeking and escaping, optimal level of arousal and peak experiences, do provide an insight into some of the factors which result in the demand for marine recreational activities. In addition, solitude and 'wilderness' experiences are sought by many visitors to the sea.

CHARACTERISTICS OF MARINE TOURISTS

It is difficult to make generalisations about the characteristics of marine tourists, because within every marine recreational activity there will be a diverse range of age groups and peoples represented. However, for particular activities a number of demographic patterns exist. For example, most of the activities which are perceived as being more 'adventurous' or having higher risk of injury, such as surfing, windsurfing, sailing and scuba, tend to be dominated by males and younger age groups. Those activities that are either passive, wildlife based or social tend to be dominated by older age groups and, in some cases, by females. An additional generalisation that applies to most marine activities is that they tend to be patronised, relative to other land-based recreational pursuits, by upper socio-economic groups. This is understandable given the often significant cost of equipment associated with marine activities. Examples include boats, scuba gear, yachts, surfboards and windsurfers. In addition, coastal areas tend to be the most desired places for homes, resorts and other development. Consequently they are more often in closer proximity to those with higher incomes, and often the fees associated with use of these areas exclude those with lower incomes.

Research on the socio-demographic characteristics of specific marine recreational activities is scarce. However, a number of studies have been undertaken. It is clear, for example, that the cruise-ship industry attracts older and relatively wealthy clients. Data from the North American market, which forms more than 80 per cent of all cruise-ship passengers, show that the mean age of passengers is 50 years and the mean income $US63,000 (Table 3.1) (Peisley, 1995). However, as we noted in Chapter 2, the demographics of cruise passengers is changing:

Table 3.1 Demographics of cruise passengers from North America 1994

Demographic category	%
Sex:	
Male	54
Female	46
Age:	
25–39	29
40–59	36
60 and over	35
Mean	50 years
Annual income ($US):	
20,000–39,000	31
40,000–59,900	30
60,000–99,900	28
100,000 and over	11
Mean	$63,000
Marital status:	
Married	76
Single	24
Type of household:	
Children in household	27
Vacation with children	15
Vacation without children	12
No children in household	72

Source: after Peisley, 1995

> Cruising is no longer the preserve of the wealthy elderly. Cruise passengers are increasingly younger and from more moderate income groups than in the past. Most (68 per cent) are married and 58 per cent travel with their spouses. Few (only 6 per cent) travel alone.
> (Caribbean Tourism Organisation, 1993: 2)

Research on whale watchers shows that they also are predominantly from higher-income and older age groups and are relatively well educated (Forestell and Kaufman, 1990; Neil, Orams and Baglioni, 1995). Interestingly, however, there is a greater proportion of females who participate in whale watching (Neil, Orams and Baglioni, 1995). This tendency has also been noted with dolphin swim participants (Amante-Helweg, 1995).

In contrast, scuba diving is dominated by males and younger age groups (Davis and Tisdell, 1995), as are activities such as sailing, windsurfing and surfing (personal observation). As would be expected, those activities that are perceived to be more adventurous and higher risk are patronised mainly by males (Ewert, 1989).

More passive activities, such as wildlife watching, beach walking, sunbathing and swimming show a much more diverse demographic profile (Anderson, 1994).

DEMAND

The demand for marine recreation and tourism can be analysed at two levels. First, specific locations have become popular settings for marine-based recreation, and second, specific marine recreational activities themselves have become popular.

Many types of marine setting are popular with tourists, but beaches are by far the most popular, particularly those in close proximity to urban areas:

> The demand for beach and bathing facilities has largely paralleled the demographic developments ... Urban beaches are increasingly seen as the single most important recreational outlet for a large segment of the urban population.
>
> (West, 1990: 263)

This demand is reflected in the massive numbers who visit beaches; for example, Miami Beach, Florida, hosts in excess of a million visitors each year. The popularity of beaches (or regions of beaches) as settings for tourism is also reflected in popular culture, as represented in music, art, movies, television and writing. Examples of such regions include Surfer's Paradise (Queensland, Australia), Copacabana (Brazil), Waikiki (Hawaii), the Riviera (France), San Sabastian (Spain), Venice (California), Acapulco (Mexico), the Golden Mile (Durban, South Africa) and Uluwatu (Bali, Indonesia). Each of these areas hosts well over a million visitors each year. This pattern of intensive use is repeated around the world at virtually every beach located close to an urban area.

While beaches are without doubt the most popular marine tourist attractions, a second important location is islands. Of course these islands also include beaches; however, they provide a base for many marine activities and have proved to be immensely popular settings for tourism and associated development, including resorts, hotels, restaurants and activity providers. Once again, many of these islands (or groups of islands) have become famous locations for marine tourism. Examples include Hawaii, Tahiti, Fiji, Bali, Catalina, San Juan's, Key West, Martinique, Aruba, Jamaica, Bermuda, Majorca, Mikinos, Cyprus, the Seychelles, Palau, the Maldives, the Canary Islands and the Galapagos. A number of smaller islands have become famous as a result of the development of a single tourist resort. Examples include Heron and Green on the Great Barrier Reef and Phuket in Thailand.

One of the reasons for this popularity is the strong positive image that small islands, beaches, coasts and the sea have. For many people the term 'relaxation' evokes images of waters gently lapping against sandy beaches. These images have tremendous power and have been used successfully to influence people's decision making regarding their use of spare time. Thus, at a general level, locations and their images are important influences on demand.

The mental image of the 'three Ss' – sun, sand and sea – is automatic for many when they think of holidays. The fact that there are now more locations, more activities and more opportunities to experience the three Ss contributes to the demand for marine recreation, which appears to be widespread and in part latent. However, the question as to why this demand exists is a difficult one to answer. Part of the answer is provided by Jones, who claimed: 'This increase in interest is fueled by a better educated public, a public that is rapidly developing an almost insatiable curiosity about the wonders of the sea' (Jones, 1993: 38).

Other environmental settings, mountains for example, offer a similarly diverse range of recreation opportunities as well as beautiful natural settings. However, the world-wide use of coastal environments for recreation far exceeds any other. Why is this the case? The answer lies in the complex number of influences on demand. Fabbri points out:

> that quite often in market-oriented economies, the demand tends to be a response to the offer. The more pressing the offer, the higher the demand, and there is no doubt that the offer of recreation and/or vacation from coastal areas exceeds by far, both in intensity and variety, offers from any other place.
>
> (Fabbri, 1990: xiv)

Fabbri's point is well made. It is important to recognise that demand for marine recreation and tourism is not a simple function of an inherently desired setting or activity. Rather it is an outcome of a complex relationship between opportunity, image, perceived benefit, cost and history. A major influence over the demand for marine tourism activities has been the invention of new technology and, through this, the creation of new activities and new locations. Thus, 'the offer' and more specifically the marketing of that 'offer' have created demand for the activity or location.

Specific activities have become attractions, irrespective of the location where they are undertaken. The invention and marketing of activities such as surfing, windsurfing, water-skiing, scuba diving and para-sailing have created a demand for these activities. It is difficult to find any marine recreational activity that is not experiencing rapid growth. This growth is indicative of the massive interest in and demand for marine-based recreation. Kenchington states that: 'Studies in Australia and elsewhere have demonstrated that fishing is the most popular participatory recreational activity. The image of leisure pleasurably anticipated is a day off to go fishing, or retiring to go fishing' (Kenchington, 1990a: 25–26). Whilst this may be true for a proportion of the population, particularly older males, marine environments have a diverse appeal for many recreational activities and there are a number of other activities that have experienced spectacular growth in participation. A good example of this is the scuba-diving industry.

Since its invention in the 1950s scuba has spawned an entire new marine tourism industry. Television programmes, such as Jacques Cousteau's *Undersea World*, helped to stimulate an interest in exploring and enjoying the underwater environment. Thus, in a pattern often repeated in marine tourism, the invention of new technology and publicity associated with its use created a 'demand' for the activity.

A number of commentators claim that scuba is amongst the fastest-growing sports in the world (Tabata, 1990; Dignam, 1990), and whilst this claim seems to be made by almost all new sports, there is no doubt that scuba has become an immensely popular activity. Davis and Tisdell (1995), who review estimates of numbers of scuba divers in Australia, claim that around 100,000 people learn to scuba dive each year there. They also point out that, in addition to certified divers, so-called 'resort dives', where non-certified divers are taken out diving under supervision, are growing rapidly. Davis and Tisdell (1995) estimated that there were around one million recreational dives undertaken annually off the Queensland (Australia) coast alone. Data from North America also reveal the popularity of scuba diving: West (1990) claims that there are between four and five million certified scuba divers in the United States. The publication of many dive magazines and dive videos, the memberships of dive clubs, the operation of 'live-aboard' dive vessels, the creation of dive-oriented resorts such as Heron Island on the Great Barrier Reef, and the establishment of many dive shops and dive tour operators all attest to the growth of scuba as a significant marine tourism activity.

A further example, which typifies the demand for marine tourism, is the cruise-ship industry. Whilst sea cruises have existed for many decades, they have predominantly been an activity only afforded by the wealthy. It is only in recent years that the industry has started to target potential customers with more modest incomes. Again, it is difficult to know whether the demand for sea cruises existed amongst the wider population anyway or whether the creation and promotion of cruise opportunities for middle income sectors created the demand. The result, however, has been spectacular. The number of cruise-ship berths (North American market) increased from around 56,000 in 1981 to 82,800 in 1990 (Caribbean Tourism Organisation, 1993). World-wide the growth has been massive:

> World demand for cruising has grown from 1.5 million passengers in 1980 to 3.5 million passengers in 1989. The decade of the 1980s recorded an average 10.3% passenger growth rate, and expectations are that the cruise industry will handle a projected 10 million passengers by the year 2000.
>
> (Marti, 1992: 360)

More recent data (Table 3.2) suggest that this estimation is optimistic. Nevertheless the industry's recent rapid growth is predicted to continue, even if at a more modest rate of around 5 per cent annually (Peisley, 1995). Associated with the forecast demand for cruises is a massive $US9 billion investment in new cruise ships by every major cruise-ship company (Peisley, 1995) (Table 3.3) including several new 'mega-ships', the largest passenger ships in history, each with capacities of 2,600 passengers (Major, 1995).

Table 3.2 United States cruise passenger growth 1989–1994

	1989	*1990*	*1991*	*1992*	*1993*	*1994*
Number of passengers (million)	3.3	3.6	4.0	4.1	4.5	4.6
Annual average growth rate	1980–1994: 9.2%			1989–1994: 7.0%		

Source: Peisley, 1995

The demand for marine recreation is immense and the growth in patronage of both locations and activities provides evidence to that effect. However, it is important to acknowledge the influence of supply, and in particular the marketing of those supply opportunities in generating that demand. While massive demand for marine tourism opportunities appears to be widespread, it is not guaranteed. A number of cases have shown a decline in tourist popularity. For example, the number of international tourist arrivals declined in south Florida during the early 1990s, primarily as a result of the negative image created by several well-publicised murders of tourists in the Miami area (Schiebler, Crofts and Hollinger, 1996). As noted in Chapter 2, tourist arrivals to Fiji dropped significantly after the military coup there in 1987 (Miller and Auyong, 1991), and the devastation caused by Hurricane Hugo, which hit many Caribbean islands in 1989, resulted in a massive downturn in tourist numbers. Tourism growth associated with marine attractions is not therefore universal.

SUMMARY

This chapter has discussed the motivations and characteristics of marine tourists. In addition, the immense demand for marine recreation, as evidenced by the massive numbers participating in marine recreational activities, has been outlined. Because the marine tourism industry is so diverse it is difficult to encapsulate any of these issues in a text of this nature. However, issues of curiosity and escape from the familiar as well as a balance between stimulation/excitement and relaxation appear to be important underlying influences over marine tourists' behaviour. In addition, a desire for feelings of competence and social interaction are also significant in determining what marine tourists do.

Table 3.3 Cruise ships under construction 1995–1998

Line	*Ship name*	*Tonnage*	*Passenger capacity*	*Cost ($USm)*
1995				
P&O Cruises	*Oriana*	69,000	1,975	355
Royal Caribbean	*Legend of the Sea*	70,000	2,068	325
Crystal Cruises	*Crystal Symphony*	48,000	1,010	250
Delta Steamboat	*American Queen*	4,700	430	55
Swedish American	*Radisson Kungsholm*	8,000	232	140
Regency Cruises	*Regent Sky*	50,000	1,400	170
Carnival	*Imagination*	70,300	2,634	330
Celebrity	*Century*	72,000	1,740	317
Princess Cruises	*Sun Princess*	77,000	1,950	295
1996				
Carnival	*Inspiration*	70,000	2,040	270
Royal Caribbean	*Splendour of the Seas*	70,000	1,800	330
Holland America	*Veendam*	55,000	1,266	225
Bergen Line	*Polarlys*		480	
Bergen Line	*Nordkapp*		490	
DSR		38,000	1,250	80
Costa Cruises	*Costa Victoria*	74,000	1,900	377
Celebrity	*Galaxy*	72,000	1,740	317
Carnival	*Destiny*	70,000	2,600	400
Royal Caribbean	*Grandeur of the Seas*	73,000	1,950	312
1997				
Royal Caribbean	*Rhapsody of the Seas*	75,000	2,000	270
Princess Cruises	*Dawn Princess*	77,000	1,950	330
DSR		38,000	1,627	225
Bergen Line	*Nordnorge*		490	
Royal Caribbean	*Enchantment of the Seas*	73,000	1,950	80
Princess Cruises	*Grand Princess*	100,000	2,600	377
Holland America		62,000	1,320	317
Costa Cruises	*Costa Magica*		2,100	400
Celebrity		70,000	1,740	317
Silversea		28,000	2,200	312
1998				
Carnival		70,000	2,040	320
Royal Caribbean	*Vision of the Seas*	75,000	2,000	275
Carnival		100,000	2,600	400
Disney Cruise Lines			2,400	
Disney Cruise Lines			2,400	
Carnival		70,000	2,040	310
Total				> 9,000

Source: after Peisley, 1995; Major, 1995

Marine tourists are diverse, and their characteristics are determined more by the nature of the activity in which they are involved than by any other influential variable. Those activities that are passive and expensive are dominated by older age groups and, understandably, by upper socio-economic groups. Activities that are active and involve elements of risk and/or physical fitness and strength, such as surfing, windsurfing and scuba diving, tend to be patronised by younger males. Interestingly, there is some evidence to suggest that wildlife interaction, with the exception of fishing which is male dominated, is more popular with females. There is little information that analyses participation in more general marine recreational activities such as beach walking, sunbathing or swimming.

There is much evidence that the demand for marine tourism is massive. This is reflected in the almost universal increase in patronage of marine tourism locations and activities. However, the demand for marine tourism is significantly influenced by the supply, and more importantly the marketing, of marine tourism opportunities. This issue is investigated further in the next chapter.

REVIEW QUESTIONS

1. Why is it useful to understand the motivations of marine tourists?
2. Select one motivation theory that applies to you. Discuss why it applies.
3. Why has fishing got little or nothing to do with catching fish? Discuss this question with regard to the complex range of motives for marine recreation.
4. Why are cruise-ship companies investing large amounts of money in building new ships?

4

THE SUPPLY OF MARINE TOURISM OPPORTUNITIES

INTRODUCTION

Whilst the amount of ocean available for marine tourism is limited, the opportunities for using it for recreation appear limitless. However, what has become apparent is that there are levels of use beyond which unacceptable levels of deterioration in resource quality or the quality of the recreational experience occur. This 'carrying capacity' idea has received much attention in the literature, particularly in the management of terrestrial park areas (Stankey, 1985). The impacts of marine recreational activities and the concept of carrying capacity are considered in the next chapter. This chapter considers the wide range of environments in which marine tourism occurs, particularly as it relates to features that attract tourists to marine settings; for while aspects of the marine environment are important components of the experience, many other factors are important in the supply of marine tourism experiences.

The 'supply' of opportunities for marine recreation can be categorised as based on activities, nature, cultural or social attractions, or special events. These categories are dependent on the prime attraction for the tourist. It is recognised that in many situations a variety of attractions exists. For example, a surfer may choose to visit Tavarua, Fiji, because it offers great surf, a beautiful natural environment and an opportunity to have a holiday with friends. Thus, the distinctions drawn here are primarily for analysing the supply of opportunities for marine tourism according to the motivation of the tourist.

Understanding the geographical spread of activities, the experiences marine tourists seek and the types of environment in which these activities occur is a confusing task. However, such matters are important if marine tourism is to be understood better. Consequently, a framework for analysing marine tourism activities and opportunities is offered in the first part of this chapter. The diverse range of recreational activities which are associated with the sea is categorised in a typology. This 'Spectrum of Marine Recreation Opportunities' attempts to clarify the influence of distance from shore, the characteristics of the environment and the experiences sought by marine tourists. The second part of the chapter considers tourism industries that are based upon specific marine recreational activities. A third section outlines the attraction of specific marine environments, and lastly the influence of social and cultural attractions and special events is considered.

THE SPECTRUM OF MARINE RECREATION OPPORTUNITIES

In 1979 Roger Clark and George Stankey of the United States Forest Service proposed a model that could be used to clarify the diverse range of recreational activities and settings that were available in forests and other large natural areas. They called this model 'the Recreation Opportunity Spectrum' (ROS) and it has become an extremely popular tool for both describing and planning for outdoor recreation (Manning, 1986). The authors, however, caution those who would apply the model as a mechanism for prescribing recreational use in the outdoors. The model's attributes lie primarily in its use as a descriptive and analytical tool. The ROS approach is to divide the natural environment up into categories based upon its physical characteristics, the recreation experiences it offers and the degree of human influence on the environment. The model has proved extremely helpful in analysing large land areas utilised for recreation (Manning, 1986).

The diverse range of opportunities for tourists to recreate in the marine environment can also be viewed as a spectrum (the Spectrum of Marine Recreation Opportunities – SMRO) and can be represented graphically (Table 4.1). The spectrum categorises marine recreation according to its distance from shore because it is this single factor that most strongly influences the activities undertaken, the experiences available and the type of environment in which activities occur. At one extreme of the spectrum, the near-shore environment, tourists are able to undertake a wide variety of shore-based activities which are easily accessible, in an environment commonly influenced by human-built structures. At the other extreme are those activities that occur far from shore, on the open ocean. These activities are based on ocean-going vessels and are usually characterised by isolation, closeness to nature and little contact with others. In between these poles lies a variety of settings and experiences which show a general pattern of decreasing social contact, decreasing human influence and decreasing numbers with greater distance from shore.

It is helpful, when analysing recreational activities available in a particular location, to simplify the wide range of opportunities so that the role of each is better understood. Consideration of where a particular marine tourism operation or activity lies in the SMRO clarifies the environmental characteristics and the experiences available for the wide variety of marine tourism enterprises.

Table 4.1 The Spectrum of Marine Recreation Opportunities

Characteristics	*Class I* *Easily accessible*	*Class II* *Accessible*	*Class III* *Less accessible*	*Class IV* *Semi-remote*	*Class V* *Remote*
Experience	Much social interaction with others High degree of services and support Usually crowded	Often contact with others	Some contact with others	Peace and quiet, close to nature Safety-rescue available Occasional contact with others	Solitude Tranquility Closeness to nature Self-sufficiency
Environment	Many human influences and structures Lower-quality natural environment	Human structures/ influences visible and close by	Few human structures close by – some visible	Evidence of some human activity, e.g. lights on shore, mooring buoys	Isolated High-quality Few human structures/influences
Locations	Close to or in urban areas Beaches and intertidal area	Intertidal ⟶ 100 metres offshore	100 metres ⟶ 1 km offshore	Isolated coasts 1–50 kms offshore	Uninhabited coastal areas > 50 kms offshore
Examples of activities	Sunbathing People watching Swimming Playing games Eating Skimboarding Sightseeing	Swimming Snorkeling Fishing Jet-skiing Non-powered boating Surfing Para-sailing Windsurfing	Usually boat-based Sailing Fishing Snorkel/scuba diving	Some scuba diving Submarining Powerboat (offshore equipped) Sailing – larger sailboats	Offshore sailing Live-aboard offshore fishing Remote coast sea-kayaking

⟵ Intensity of use

⟵ Human impact

ACTIVITY-BASED MARINE TOURISM

There are many activities that have become universally popular in marine situations. Whilst they are dependent on certain types of marine environment or condition, the prime attraction for the participants is the activity rather than the location. Examples include surfing, windsurfing, fishing, scuba diving, water-skiing and sailing. Each of these activities has millions of regular participants. With the exception of scuba diving (which does not have a competitive element), each has both competitive and recreational aspects and in many situations a professional sporting competition has developed. In addition, each of these sports has developed an image, or series of images, and what could be termed a 'culture'. Surfing, for example, has spawned a multimillion dollar clothing industry with labels such as Rip-Curl, Billabong, Stussy, Hang Ten, Ocean Pacific, Lightning Bolt, Bear, Town and Country, Mambo and Rusty. Popular bands also utilise and contribute to the surfing image; examples include the Beach Boys, the Hoodoo Gurus, the Butthole Surfers, Australian Crawl and Mental as Anything. In addition, movies such as *Endless Summer*, *Big Wednesday* and *Point Break* and television shows like *Bay Watch* and *Paradise Beach* reflect the 'lifestyle' that has developed around the activity of surfing. Thus, marine recreational activities have an influence that extends far beyond an enthusiasm for the activity itself. For many participants, and even for non-participants, the image and lifestyle associated with the activities are attractive and become a part of 'who they are'.

The supply of opportunities for participating in these kinds of activities is largely dependent on the conditions needed for each. For example, water-skiing is dependent on relatively calm water-surface conditions, while of course surfing is entirely dependent on the opposite. Many activities, however, have diverse forms adapted to differing conditions. For example, sailing occurs both close to shore (SMRO class II) in small vessels, near shore (SMRO class III) and offshore (SMRO class V). Similarly, fishing occurs in all SMRO classes, although its occurrence offshore (SMRO class V) is limited.

An excellent example of a rapidly growing marine recreational activity that has proved controversial is that of personal watercraft.

Case study: personal watercraft

Growth in the use of jet-skis, wet bikes and water scooters, generically referred to as personal watercraft (PWC), has been spectacular. They represent one of the fastest-growing marine recreational activities and their use has resulted in controversy.

The majority of PWC share a number of common features. They are small in size, contain powerful motors and have a shallow draft that allows them to be used in water close to shore. In the early 1990s it was estimated that there were approximately 400,000 PWC in the United States and that the annual growth rate was of the order of 100,000 units per year (Cuthbert and Suman, 1995). Other countries such as New Zealand and Australia also appear to be experiencing rapid growth rates. However, exact figures are virtually impossible to establish.

This growth can be attributed to a range of factors. Engine manufacturers in Japan, such as Suzuki and Kawasaki, are increasingly diversifying and pursuing new markets in order to increase profits (Bairstow, 1986). As PWC are being manufactured at increased rates by such companies they have decreased in price, making them more affordable. Engine refinements have also made PWC more powerful, with speeds of over 80 km per hour easily attainable. A further feature contributing to their popularity is that they are relatively simple to ride at high speeds. Very little knowledge or training is required to operate them. However, novice operators have little understanding of distances, speeds and manoeuvring, and often have poor control of the craft.

Although PWC are undoubtedly fun for the rider, they have become increasingly controversial. They are a direct threat to slow-moving marine life such as turtles and manatees. In Florida, significant numbers of manatees are struck by small vessels, including PWC, each year (O'Shea, 1995). The noise from PWC may also drive nesting birds away from breeding areas (Cuthbert and Suman, 1995). The shallow draft of PWC allows them to operate in estuaries and mangrove forests and over sand flats that are often important feeding and breeding grounds for wildlife.

The impact PWC are having on humans is also becoming increasingly clear. Conflict has grown between PWC operators and other marine user groups. Many anglers, for example, are opposed to PWC because they disturb fish life. Kayakers and sailors are complaining about the noise pollution emanating from PWC and the dangerous actions of some riders (Cuthbert and Suman, 1995). Noise pollution complaints are also voiced by those ashore. As well, PWC pose a direct threat to the physical safety of other marine users: these craft have killed kayakers, swimmers and other PWC riders in New Zealand, Australia and the United States.

PWC have also been identified as a potential threat to tourism, as they damage the perception of tranquil environment by those tourists not participating in their use (Cuthbert and Suman, 1995). Evidence from Kaneohe Bay, Hawaii, suggests that tourists inexperienced in the use of PWC have difficulty doing even basic manoeuvres such as turning and stopping (Mattix and Goody, 1990). This suggests that inexperienced users may be an even greater danger to themselves and other marine users.

NATURE-BASED MARINE TOURISM

There is widespread agreement that the demand for nature-based tourism is increasing rapidly (Orams, 1995). An argument can be made that, as a result of pollution and the destruction of natural ecosystems, the supply of opportunities for high-quality nature-based recreation is decreasing. Comparatively, however, the supply of such opportunities in the marine environment is actually increasing. This is a result of the increasing accessibility of high-quality marine ecosystems. Evidence has been presented earlier in this book that new technology and greater marketing of that technology has greatly expanded the accessibility of the marine environment. Thus, the availability of high-quality nature-based experiences has increased for tourists. However, it must be recognised that the actual number of high-quality marine ecosystems is limited and is decreasing. Consequently, there will come a time in the future where the possibilities for experiencing such settings will also decrease.

There are a number of cases where the 'explorer tourists' of the marine world have 'discovered' pristine new locations and have provided the impetus for the development of those locations for tourism. The classic progression through Butler's 'destination life cycle model' (Butler, 1980) has resulted and thus the natural qualities of the site have been degraded. The degradation of natural attractions as a result of tourism is a widespread phenomenon. These impacts are discussed in more detail in Chapter 5. Here a case study which indicates the attraction of nature and the impacts of tourism based upon it is that of Australia's Heron Island, Queensland.

Case study: Heron Island

Heron Island is located in the southern region of Australia's Great Barrier Reef amongst a group of around 13 islands known as the Capricorn-Bunker Group. Heron is a sand cay located at the western end of a coral-reef lagoon and is relatively small, at around 800 m long and 280 m wide at its widest point. The island is an ecologically important breeding site for birds such as the black noddy (*Anous minutus*) and the wedge-tailed shearwater (*Puffinus pacificus*) and also for a number of species of sea turtles, especially the green sea turtle (*Chelonia mydas*).

Early human use of the island was based on the harvesting of turtles, and a turtle-soup factory operated at Heron from 1925 to 1929 (Limpus, Fleay and Guinea, 1984). This factory became a tourist resort in 1932 (Great Barrier Reef Committee, 1977) and it has been subsequently bought and developed by P&O. Additional development on the island includes a research station established in 1951, now operated by the University of Queensland (Jones, 1967), and a park centre for Heron Island National Park established in 1983 (Neil, 1993).

Tourist use of Heron Island has changed significantly since the mid-1980s. Hopley (1988) suggests that two major factors were responsible. First, the introduction of large, high-speed catamarans facilitated rapid, relatively comfortable access for tourists. Heron is now only two hours away from the mainland, whereas prior to the introduction of the catamarans, boat trips could take up to six hours. Second, the variety of activities available for tourists has significantly increased. Scuba diving, snorkelling, semi-submersibles and glass-bottom boats have all 'opened up' the underwater world for visitors to Heron. The result has been a larger number of tourists who visit more of the environment surrounding Heron. The resort is estimated to host about 11,000 tourists annually. It is, therefore, the second most intensively used coral cay on the Great Barrier Reef and the most intensively used in the southern reef region (Heatwole and Walker, 1989).

Heron Island Resort promotes itself as an 'environmentally friendly' nature-based tourist facility:

> Whilst the island is an international resort, great care is taken to make it a 'live and let live' situation with nature. The resort takes up only one corner of the island, with the remainder belonging to the seabirds and the turtles ...
>
> The golden rule of Heron is to enjoy nature, without disturbing it. This is carried out right through to details like garbage disposal. Resort waste that is not biodegradable, such as bottles and cans, is actually shipped back to the mainland and properly disposed of. Every precaution is taken so as not to upset the delicate balance of nature.
>
> (Heron Island Resort, 1986: 2)

However, despite these 'green' sentiments there is no doubt that the development and operation of the resort at Heron Island has caused significant detrimental environmental impacts. It should also be noted that the operation of the University of Queensland research station and the national park centre also contributes to these impacts.

Examples of these impacts include the building of structures resulting in a reduction in the area available for nesting for both sea birds and turtles. This is particularly significant for the green sea turtle and the burrow-nesting wedge-tailed shearwater, both of which are thought to return to the same site each year for nesting. The lights from buildings are also thought to confuse hatchlings from both these species, resulting in increased mortality rates (Gibson, 1976; Hulsman, 1983; Lane, 1991). Tourists who walk around the island have been observed to collapse the shearwater burrows, killing the nesting bird and chick (Dyer, 1992). In addition, disturbance of sea birds can cause them to abandon their nests, and hatchlings can die if the parent does not return after as little as 15 minutes (Hulsman, 1984).

Of more direct concern with regard to the marine environment is the dredging and maintenance of the boat channel through the reef flat to the western side of the island. This channel has significantly altered the tidal flow around the island and the surrounding reef flat (Gourlay, 1991), altering the benthic fauna adjacent to the channel (Neil, 1988) and accelerating beach erosion on the northern and southern sides of the island (Neil, 1993).

Tourists and their actions also impact on Heron's ecology. The common practice of walking over the coral reef flat at low tide at Heron can significantly reduce the health and abundance of coral (Woodland and Hooper, 1977). Litter is common on both the inter-tidal areas and the reef flat (personal observation). Recreational fishing is also likely to alter the natural composition of reef fish populations.

Heron Island provides an interesting example of a marine tourist destination that offers and markets itself as a nature-based attraction. However, despite the best intentions of minimising impacts, they do still occur. This pattern is common amongst most marine tourism destinations.

SOCIAL AND CULTURAL ATTRACTIONS

It is well documented that social interactions form an extremely important part of tourism experiences (Leiper, 1995). Opportunities to holiday and have fun with friends remain one of the most important motivators for recreational travel (Leiper, 1995). Consequently, the supply of such opportunities is an important issue to consider in a discussion about marine tourism. While it may be contentious to classify an attraction like 'Muscle Beach' (a body-builders' weight-lifting area) at Venice Beach, California, as a 'cultural attraction', there is no doubt it is a tourist attraction that focuses on human diversity. In fact, 'people watching' is an extremely important part of most popular beach areas. Many tourists visit locations such as Miami Beach to look at scantily clad men and women because they are different from people at home. For many teenagers around the world, summer-time trips to the beach are primarily motivated by opportunities to investigate members of the opposite sex with few clothes on.

Marine settings are also locations that are used for social gatherings and celebrations. Bonfires on beaches, picnics and barbecues, the 'rafting up' of boats for socialising, parties on cruise boats and beach-based recreational team sports are all examples of the importance of social interaction in many marine tourism activities.

In more traditionally 'cultural' terms, historical events and museums are also important marine tourism attractions.

Case study: maritime museums and historical festivals

Maritime museums are found in almost every country that has links to the sea. These museums not only preserve the socio-cultural history of a country's maritime past but also serve as marine tourism attractions themselves. The vast majority of such museums are located alongside harbours or waterways. Many of the attractions themselves, such as the sailing ship *Peking* at the South Street Seaport Museum in New York and the *Great Britain II* in Bristol, England, are still afloat, being permanently moored as walk-aboard displays. Other exhibits, such as the 'coastal traders' at the Auckland Maritime Museum in New Zealand, are operational and take tourists on harbour excursions.

Maritime museums the world over are adopting new strategies to attract tourists. The Deutsches Schiffahrts museum in Bremerhaven, Germany, offers guided tours, evening lectures and educational programmes. In addition, the museum has been developed to allow tourists to participate in activities such as manoeuvring scale models by remote control and operating a submarine periscope (Ellmers, 1991). Increasing numbers of museums are adopting such interactive approaches to displays.

Other museums, such as the Viking Ship Museum in Roskilde, Denmark, are entirely devoted to preserving and exhibiting one period of marine history. This museum, as its name implies, focuses on Viking ships and artefacts. The museum was constructed in the late 1960s to house five Viking ships that had been unearthed in the adjacent fjord.

The museum is a 'living workshop' at which archaeologists continue to undertake restoration work (Madsen, 1991). This work is often undertaken in public viewing rooms to allow visitors to understand the restoration process. In addition, audiovisual displays illustrate the recovery of the original five Viking ships. The museum has also built replica Viking ships using traditional methods. These building activities are on display too (Madsen, 1991).

The South Street Seaport Museum is one of the world's most active maritime museums. The museum operates 'excursion vessels' which in the early 1990s were taking more than 300,000 passengers annually around the harbour (Neill, 1991). The two schooners *Lettie G. Howard* (1893) and *Pioneer* (1885) are also operated as working historic vessels. Tourists allow the boats to earn the cost of their operation and maintenance while providing an enjoyable activity (Neill, 1991). The museum is also very active in organising marine festivals. In 1986, Operation Sail, a festival to celebrate the Statue of Liberty Centennial, gathered together tall ships from 23 countries (Neill, 1991). Over the three-day event, 40,000 spectator vessels attended, while on shore 350,000 people visited the 15 historic ships docked at the museum.

SPECIAL EVENTS

There is a multitude of specific marine-based events that are significant tourist attractions. Many of these events are regular occurrences, such as those that are held on an annual basis. Examples include competitive marine sporting events such as national surf life-saving championships, national yachting regattas and annual fishing competitions. Others have a recreational focus – for example, an annual mid-winter swim – or an environmental theme, like an annual beach clean-up. Many of these events attract tens of thousands of spectators in addition to the participants. Consequently an event can have major economic implications for the surrounding community. For example, a study conducted on the impacts of the 1997 Billabong Pro surfing event held on Queensland's Gold Coast, Australia, found that 32,000 spectators attended the five-day competition and that the net economic impact of the event was around $A2.3 million (Kavanagh, 1997).

The great majority of special events occur in SMRO class I and II settings. However, there are some events, usually off-shore yacht races like the Whitbread Round the World Yacht Race, or the Trans Atlantic, that are conducted primarily in SMRO class V. Probably the most significant marine tourism event in terms of longevity, spectator interest and economic influence is the America's Cup.

Plate 4.1 Special events such as the Whitbread Round the World Yacht Race have become significant tourist attractions.

Case study: the America's Cup

The America's Cup is promoted as the oldest international sporting event. In 1853, after an invitation was extended to the United States to compete in a yachting regatta off the south coast of England, the yacht *America* defeated the entire English sailing fleet. The trophy that was awarded to the winning crew became known as the America's Cup. It has subsequently become a trophy that yacht clubs from different countries compete for. The cup holder 'defends' in a series of races against the top international challenger. The winner of this America's Cup match is awarded the cup and must defend it against subsequent challengers.

The unique history of the America's Cup is that the United States successfully defended it for over 130 years. Finally, in 1983, the Australian challenger defeated the US defender 4–3 and the cup went to Australia. Australia conducted the defence of the America's Cup in the city of Fremantle in 1987, where the US challenger *Stars and Stripes* defeated the Australian defender *Kookabura III* 4–0, and the cup returned to the United States. The United States successfully defended in 1992 but lost the cup to New Zealand in 1995. The next America's Cup will be held on the waters of the Waitemata Harbour in Auckland New Zealand in the summer of 1999–2000.

The America's Cup event in New Zealand will include five months of competition between the challengers (a record 18 challenges have been received) and over 250 races. The final America's Cup match (best of nine races) is predicted to attract over 6,000 spectator vessels, around 70 helicopters overhead, over 2,000 journalists and around 20 television crews. Over 160 countries will take television coverage of the event (Thomas, pers. comm.).

Initial studies predict that the America's Cup event will have important economic implications for Auckland. The cup is predicted to increase tourism numbers to the area by 5 per cent (60,000 additional visitors). These additional visitors to the region are expected to spend over $NZ110 million. Additional expenditure from challenging syndicates is predicted to exceed $NZ80 million. Further expenditure in the region will result from infrastructural developments designed to support the cup event. In total the event is predicted to contribute around $NZ466 million (Brent Wheeler and Co., 1997).

The America's Cup is a major marine tourism event that has a long history and has grown in influence and impact in recent years. It will be the largest sporting event ever hosted in New Zealand and provides that country with an important event from which it can 'lever' and market its products and industries.

SUMMARY

The supply of marine tourism opportunities is closely linked with the issue of access. New inventions are creating new activities and allowing access to previously unused areas. Thus, it could be argued that the supply of opportunities for marine recreation is ever increasing. There are certainly many marine activities available now which were not available 30 years ago. However, an important issue with regard to the supply of opportunities for marine recreation is that of environmental quality. Most marine tourism activities are dependent on the quality of the resource; for example, fishing cannot occur if there are no fish. An even more extreme scenario would see a complete loss of marine recreation opportunities if an area is so polluted it is harmful to human health. Unfortunately, this is a reality for the harbour and beach areas close to large cities. Thus, the supply of marine activities, while increasingly diverse, is constrained by environmental quality. The impact of tourism and other human activities on the marine environment inevitably affects our ability to utilise that environment for recreation. The issues of impacts and management are therefore critical to the future of marine tourism. They are discussed in the following chapters.

REVIEW QUESTIONS

1. Discuss how the demand and supply of marine tourism are linked.
2. Critique the Spectrum of Marine Recreation Opportunities. Is it an accurate model? How does it help an understanding of marine tourism?
3. Discuss how nature-based marine tourism may become a catalyst for further tourism development, and consider the implications of that development for the tourism attraction.
4. Do you agree that environmental quality is an important constraint on the supply of marine tourism opportunities? If so, why do you agree? If not, why not?

5

IMPACTS OF MARINE TOURISM

INTRODUCTION

In the early stages of mass tourism development, the impacts of tourism were largely viewed as positive, particularly with regard to their influence on the economic development of a region or country. The influx of 'foreign' cash into an economy as a consequence of tourism was viewed as an 'export' industry from an economic perspective. Consequently, tourism was seen as a desirable sector to target for development. However, more recently it has been recognised that there are costs associated with tourism development:

> Countries are encouraged to invest in tourism because of its evident economic benefits – particularly foreign exchange earnings, employment and infrastructural development such as transport networks. Cost-benefit analysis of tourism development has tended to concentrate on these positive outcomes, while scant attention has been given to the social, environmental and other costs associated with development.
>
> (Warren and Taylor, 1994: 1)

Other analysts are far more dramatic in their assessment of the impacts of tourism. For example, Croall states:

> A spectre is haunting our planet: the spectre of tourism. It's said that travel broadens the mind. Today, in the modern guise of tourism, it can also ruin landscapes, destroy communities, pollute the air and water, trivialise cultures, bring about uniformity, and generally contribute to the continuing degradation of life on our planet.
>
> (Croall, 1995: 1)

While this may be a little extreme, it is indicative of the growing view that tourism is not the panacea it was once, in some cases, made out to be. It is now widely understood that there are many negative impacts that result from it.

Similarly, as the growth of marine tourism has become widespread, an increasing number of reports show that significant environmental, social, cultural and even economic damage can result. However, there are also success stories, where the development of marine tourism has, on balance, improved things. The question arises, therefore, of what it is that makes the difference between successful and unsuccessful marine tourism. Related to this first question is a second: how do we decide whether the impacts of the tourism have been positive or negative? These questions have also been asked with regard to the wider tourism industry. For example, Johnston asks:

> Is tourism development compatible with the ideals of 'sustainable' development? Can tourism, an industry that inherently creates dependency relationships, truly be tailored in a socially responsible and environmentally viable fashion?
>
> (Johnston, 1990b: 2)

The questions are difficult to answer. Considerable debate surrounds the assessment of impacts of tourism. However, the first step in such an assessment is to describe the range of impacts that tourism activities and associated development are having. This chapter outlines some of these impacts, both positive and negative, and the following chapter considers strategies to manage these impacts.

ENVIRONMENTAL IMPACTS

As outlined in Chapter 4, there appears to be a consensus in the literature on tourism that demand for opportunities to interact with nature, including marine environments, has been increasing rapidly (Jenner and Smith, 1992). This general interest in nature, marine settings and experiences based upon them is reflected in an increasing demand and value being placed on relatively undisturbed coastal environments and, in particular, wild animals (Gauthier, 1993). Whilst tourism of this type has been applauded by many as a suitable saviour for threatened wildlife populations, including marine species (Davies, 1990; Groom, Podolsky and Munn, 1991; Borge, Nelson, Leitch and Leistritz, 1991; Barnes, Burgess and Pearce, 1992; Burnie, 1994), many authors are cautious regarding the negative environmental impacts resulting from nature-based tourism (Butler, 1990; Wheeller, 1991, 1994; Zell, 1992; Pleumarom, 1993). Glasson, Godfrey and Goodey summarise these views when they state: 'Tourism contains the seeds of its own destruction; tourism can kill tourism, destroying the very environmental attractions which visitors come to a location to experience' (Glasson, Godfrey and Goodey, 1995: 27).

There are significant numbers of cases which illustrate negative impacts associated with tourist–nature interaction (Hanna and Wells, 1992; Burger and Gochfield, 1993; Griffiths and Van Schaik, 1993; Ingold *et al.*, 1993; Viskovic, 1993; Muir, 1993). More specifically, there are many authors who are expressing concern over the negative impacts that are being inflicted on marine ecosystems as a result of marine tourism activities (Hegerl, 1984; Mellor, 1990; Ward, 1990; Laycock, 1991).

Several examples illustrate these concerns. In Florida the endangered West Indian manatee has become a major tourist attraction:

> This virtual certainty of seeing manatees between November–March in the Blue Springs/Crystal River area has contributed to the increased popularity of manatee-related diving and boating excursions. Within the Crystal River area alone (home to a third of the west coast Florida manatee population), five dive shops exist to service manatee-related scuba and snorkelling. In the height of the manatee season, the density of divers in the estuarine waters can reach 1/10 m^2 and, despite manatee harassment legislation, many will pet, stroke and attempt to ride the animals.
>
> (Shackley, 1990: 313)

Shackley concludes her review of manatee-related marine tourism thus: 'Anyone who wants to ensure the survival of the species would be well advised to avoid visiting them' (Shackley, 1990: 316).

A further report identifies the concern over the impact of marine tourism growth in the northern Pacific:

> Glacier Bay in southeastern Alaska has been the site of a long running controversy concerning the sensitivity of humpback whales to ship disturbance ... In 1970 only four 'large' ships (meaning mainly cruise ships but occasionally also meaning state ferries and military vessels) entered Glacier Bay. Seven years later, 103 large-ship entries were recorded by the National Park Service, and many additional visits were made by smaller tour vessels and private craft. A 'sudden departure' of humpbacks from Glacier Bay was reported in the summer of 1978, and again the following year fewer whales entered and remained in the bay for the summer feeding season.
>
> (Reeves, 1992: 8)

An important concept that illustrates the danger posed by small incremental changes to natural environments is 'recreational succession'. This concept was first proposed by Stankey (1985) when he described the gradual deterioration of a camping site as it became increasingly popular with

Plate 5.1 A New Zealand fur seal pup killed as a result of entanglement in a fishing net. Examples of human impacts on the marine environment like this one are widespread.

visitors. This phenomenon has been observed in many different natural settings and can be described as follows. As pristine natural sites are discovered and used for recreation, deterioration of the site's natural attributes occurs. Consequently, initial visitors, who were attracted by the pristine unspoilt surrounds, move on and are replaced by greater numbers, with lower expectations of environmental quality. This chain continues, resulting in ever-increasing numbers of visitors, increasing development of the site's infrastructure to cope with visitors' needs and decreasing environmental quality. Meantime, the initial 'discovering group', having moved on, have explored and 'discovered' another pristine site and thus have started the chain of recreational succession again elsewhere. The overall result of recreational succession is a gradual 'creep' of development of facilities and infrastructure and a gradual loss of 'wilderness' and environmental quality.

Related to the concept of recreational succession is Butler's (1980) Tourism Life Cycle model (Figure 5.1). He argues that a tourism destination or attraction passes through a number of general stages as it becomes more popular. This model reflects the typical 'product life cycle' concept which is widely used in the marketing and business management areas; that is, that a product will follow a birth–growth–maturity development pattern. This kind of general development path has been observed in long-established whale-watching locations (Forestell and Kaufman, 1995) and has relevance

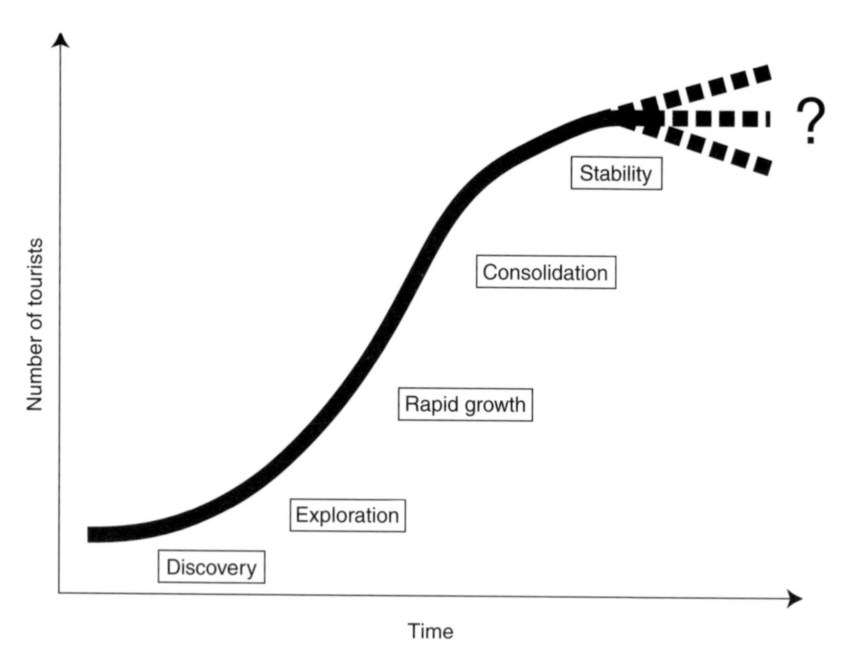

Figure 5.1 Possible stages in the development of marine tourism attractions (adapted from Butler, 1980)

to a number of marine tourism cases. For example, the case of Hanauma Bay on the island of Oahu in Hawaii illustrates both the dramatic environmental impacts that can occur as a result of tourism and the developmental stages outlined in Butler's model.

Case study: Hawaii's Hanauma Bay

Hanauma is a sheltered cove located on the island of Oahu, Hawaii, about 15 minutes' drive from the city of Honolulu. Prior to the 1950s this secluded bay was used by local Hawaiians for traditional food gathering and by occasional fishers and weekend campers. Coral, fish and other marine life were plentiful at this time and use of these resources was small-scale. During the 1950s a beach road was constructed from Honolulu to Hanauma. This improved access dramatically changed the numbers of people visiting the bay – a massive increase in use occurred over the next decade (Burgett, 1990). By 1964 it was estimated that over 1,000 fish and 500 coral heads were being removed each year by visitors to the bay using scuba and spearguns. Concern over these impacts resulted in a ban on the taking of all marine life from the bay in 1967, and in 1970 Hanauma was declared a marine park and promoted as a tourist destination (Reynolds, 1990).

A number of private tour operators began to run bus tours to the bay for tourists, for whom snorkelling and fish feeding became popular activities. The increased use of the area resulted in demands for improved facilities to serve the tourists. The road was improved, parking lots enlarged, restrooms and picnic facilities added. These facilities served to increase the popularity of the bay, and concerns over the 'carrying capacity' of the location were expressed. Local government funded a study in 1977 to determine the optimum level of use for the park. This study concluded that the carrying capacity of the bay should be set at 1,000 visitors per day. The popularity of Hanauma has continued, however, and use of the bay exceeds 10,000 visitors each day (Burgett, 1990). Reynolds (1990) estimated that by 1981 over 2 million tourists visited the small bay each year.

Assessing the impacts of this intense recreational use of the bay has been difficult. As with all marine ecosystems, the interactions of reef fish and other marine fauna with other components of the ecosystem are complex, dynamic and consequently difficult to measure. However, a number of studies have shown that the biomass (weight) of fish inside the reef crest at Hanauma is very much higher than a natural level, while the biomass of other organisms such as coral, sponges and other marine fauna is much lower and declining. Several causes are suggested. First, the common practice of fish feeding by tourists is encouraging larger than normal concentrations of reef fish species which will accept human-provided food. Second, direct trampling on benthic organisms by waders, swimmers and snorkellers is destructive. Third, the great amount of silt stirred up by large numbers of people wading in the shallows is harmful to marine life such as corals. Fourth, the suntan lotion and urine entering the water as a result of large numbers of tourists is harmful. Reynolds states that 'by ten o'clock the water is cloudy (from silt) and there is an oil slick of sun tan lotion on the water' (Reynolds, 1990: 106). Lastly, the freshwater showers installed for tourist convenience on the shore are resulting in higher than normal levels of freshwater run-off, altering the composition of the near-shore salinity.

The consensus with regard to Hanauma is that the rapid growth of tourist numbers, together with the infrastructure established to service their needs, has produced a severe reduction in the quality of the bay's ecosystem, particularly in the heavily used shallow areas. Hanauma, however, remains popular and provides an example of the concept of recreational succession. Many now consider Hanauma to be nothing more than a 'sacrifice area' – a location where the mass tourists can be channelled to concentrate their negative effects, thereby reducing the pressure on other bays and beaches on the island of Oahu.

The case of Hanauma is typical of many coastal locations that have become popular with tourists. It appears that environmental degradation is inevitable when tourism becomes established. However, despite all these 'doom and gloom' examples, there are cases where tourism development has provided the impetus for an improved local environment. For example, one of the major justifications used in the establishment of marine protected areas has been their value as tourist attractions (Salm and Clark, 1989). Similarly, the value of endangered species alive, as tourist attractions, rather than dead for food products, has provided a justification for the protection of marine animals such as whales, dolphins, seals, sea turtles and sharks (Orams and Forestell, 1995). Furthermore, there is evidence to suggest that marine tourism experiences can be utilised to prompt tourists to become more environmentally responsible – to become active marine conservationists (Orams, 1997). Environmental organisations

Plate 5.2 Massive coastal developments for tourism, such as this one in the Bahamas, have major impacts on small local communities.

such as Earthwatch, World Wide Fund for Nature, the Pacific Whale Foundation, Marine Mammal Stranding networks and many others all utilise marine tourism operations and experiences to solicit members and support for their conservation work. And, as a consequence, marine tourists do, in some cases, actively support the improvement of marine environments.

SOCIO-CULTURAL IMPACTS

In 1870 the Reverend Francis Kilvert wrote in his diary 'of all the noxious animals, the most noxious is the tourist' (Croall, 1995: 21). These sentiments are now being echoed around the globe by local people whose societies have come to be dominated by tourism. For example, Hawaiian activist Puhipau is reported to have said: 'I beg you, please don't come to Hawaii. Tourism is killing us, it is literally sucking the life out of us' (Puhipau, 1994: 10). His comment illustrates one extreme of the socio-cultural impact of tourism. There are now many nations and regions where the numbers of tourists far outnumber the locals and where the area's development, activities, employment and services are so dominated by tourism that the integrity and traditions of the local culture are completely subsumed. This kind of influence over the lives of locals by the presence of 'outsiders' produces widespread resentment.

The impact of tourism development on local communities has been characterised in a model proposed by Doxey (1975). He argues that the reactions of a host community to the growth of tourism vary over time in relation to developmental stages. The reactions range from initial cynicism and euphoria, as locals consider the possibilities for their location, to increasing levels of negative responses, as the costs of tourism development to the local community are felt (Figure 5.2). Eventually a stage of acceptance and/or adaptation to the changes induced by tourism is reached.

Forestell and Kaufman (1995) provide an analysis of the development of whale watching from their experiences in Hawaii and Australia. Their work suggests that the operators involved in whale watching experience phases of discovery early on, competition as growth develops, confrontation as regulatory agencies become involved and eventual stability when the industry matures. These observations can be considered in conjunction with Butler's life-cycle approach (Figure 5.3).

The socio-cultural impacts of marine tourism vary over time. Whilst models such as those briefly reviewed here may not have universal applicability, they emphasise that any socio-cultural assessment should consider the stage of development of a particular location. A good example of socio-cultural impacts of tourism, the majority of which is marine based, is that of the islands of the Caribbean.

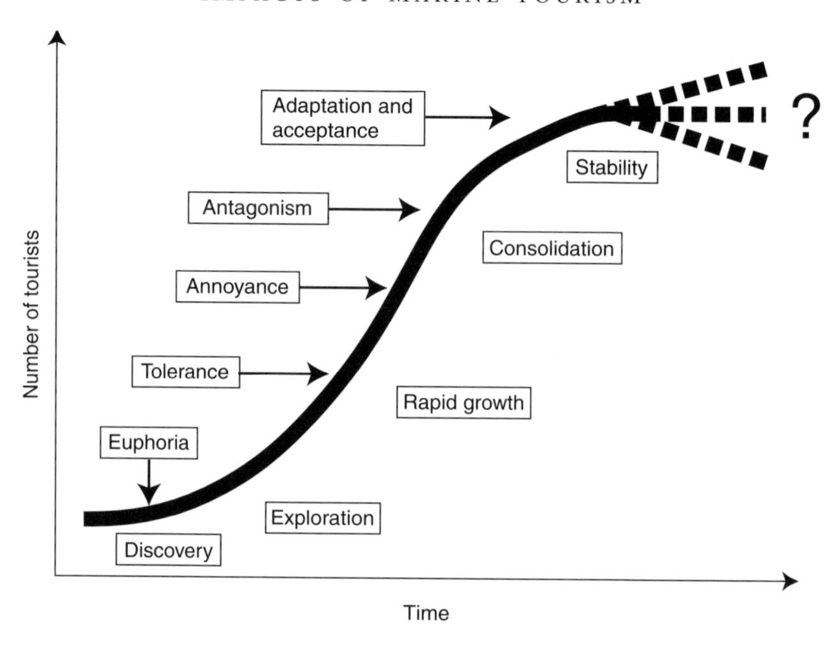

Figure 5.2 Possible stages in the development of marine tourism attractions and associated community reactions (adapted from Butler, 1980; Doxey, 1975)

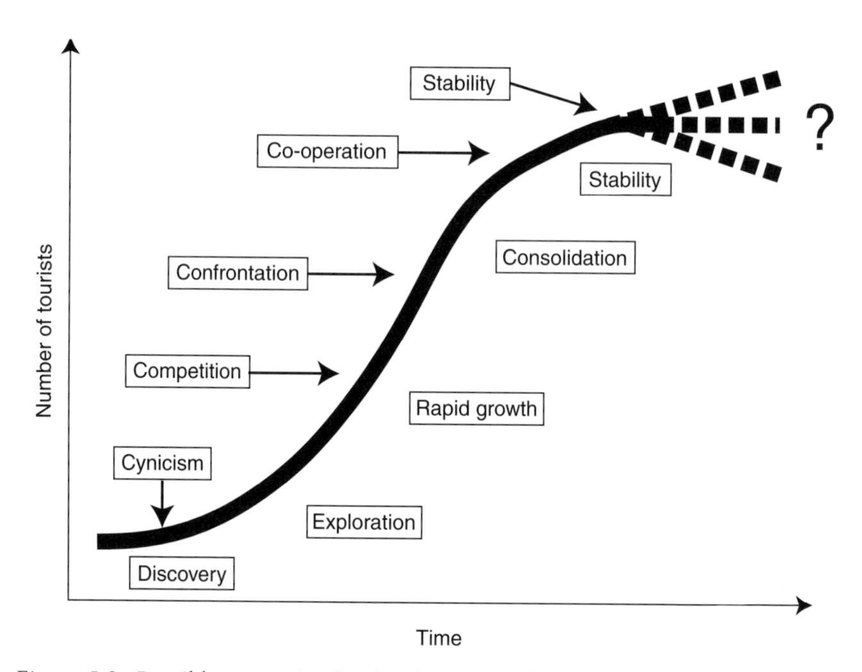

Figure 5.3 Possible stages in the development of whale-watching locations and associated operator reactions (adapted from Forestell and Kaufman, 1995)

Case study: the islands of the Caribbean

Small island nations, such as those in the Caribbean, where marine-based tourism is predominant, are particularly vulnerable to its socio-cultural impacts. The small size of these islands and their popularity result in exacerbated effects on more traditional land uses. For example, the forced displacement of the local population from favoured areas, particularly beaches, is common in the Caribbean. Prior to the development of tourism on the island of St Thomas, more than 50 beaches were available to the local population for recreation and other uses. However, by 1970 only two beaches remained open to the public; the remainder are reserved exclusively for the use of tourists. This has resulted in widespread resentment amongst the local population (Johnston, 1990a). In 1984 one of these two remaining public beaches, Magens Bay, was re-zoned by the island's government for a tourism development. This caused a massive public outcry, which resulted in the government being voted out of office (Johnston, 1990a). During this protest many strongly worded anti-tourism murals, slogans and graffitti appeared around the island.

The creation of exclusive beach clubs and resorts on Antigua has also stopped the great majority of Antiguans from using the island's best beaches. Access for locals is only possible by purchasing expensive day passes that are beyond the budgets of most.

Displacement from traditionally popular areas also occurs when older islanders feel uncomfortable with large numbers of tourists who fail to observe local dress codes and customs (Pattullo, 1996). Because few tourists visit there, Friars Bay in Antigua has become a 'refuge' for locals seeking to escape tourists and partake in traditional beach activities. A dramatic illustration of the exclusivity of beaches controlled by resorts and clubs on the island is given by Pattullo (1996), who states that the prime minister of Antigua was refused entry to a beach on the island in 1994.

Tourism in the Caribbean has also been blamed for changing the resident population's ethics and general outlook on life. A transition away from old traditional values to a 'western-style materialism' is often cited as an example of this. An Antiguan politician observed that too many islanders were 'imitating the life-style of the holidayers whom [they serve]' (Pattullo, 1996: 85). This loss of tradition or 'cultural identity' is a widely reported concern in the Caribbean (Johnston, 1990a).

A further example of the socio-cultural impact of tourism in the Caribbean has been a significant increase in prostitution and crime (Mathieson and Wall, 1982). In Antigua 'beach boys' operate on island beaches looking for western women in search of local boyfriends. Payment for their 'services' usually takes the form of meals out in restaurants, clothes and other material goods, as well as money (Pattullo, 1996). Female prostitution is also well established on Caribbean islands. In the past some marketing strategies used by Caribbean nations have even tried to cash in deliberately on the sex image (Mathieson and Wall, 1982). Tourism has created a demand for drugs, which are supplied by local dealers; and drug traffickers often use tourism to mask their activities (Johnston, 1990a). Affluent tourists have also provided local street criminals with desirable targets.

ECONOMIC IMPACTS

There is no doubt that tourism development has had widespread economic benefits. There are numerous examples of communities, regions and even nations that have been rejuvenated economically, with the resultant social benefits of greater employment, better services, improved health and generally improved standards of living (see Kaikoura case study below). While some locals may resent the prevalence of tourists in locations like Antigua or Hawaii, there is no denying that these same locals benefit from the businesses, services and infrastructure that these tourists support. The reality is that it is difficult to have the good (for example, employment opportunities and improved social services) without the bad (for example, increased traffic and inflationary pressures).

The economic influence of tourism is both pervasive and seductive. It appears to be an industry that needs little capital investment when, for example, the attractions of an area already exist. It brings a significant influx of cash into an area, spread over a wide variety of service industries such as food, accommodation and transport as well as the attraction itself. In addition, tourists contribute to the taxation revenues of governments (for example, through sales taxes on goods and services) and yet the government does not have to cater for these people in the longer term, as they are visitors. Furthermore, these visitors have little power or interest in the election of governments. As a consequence tourism appears an extremely attractive option when governments are considering how best to develop their economies. However, there are a number of cases that illustrate that negative economic consequences can result from tourism developments.

It is self-evident that if the tourism activity destroys the attraction upon which it is based then the investment in tourism infrastructure and businesses is lost. Sustainability is, therefore, a critical component of the long-term economic success of any nature-based tourism venture. This is widely recognised in marine-tourism-based communities. For example, in Western Australia at Monkey Mia it is understood that the economic benefits that accrue from the dolphins mean that it is in the local community's interest to protect the dolphins:

> Many locals are quick to point out the need for careful management of Monkey Mia to preserve the dolphin–human interaction and the windfall of tourist dollars for Denham, the commercial centre of the Shire.
>
> (Keys, 1987: 23)

One of the most common economic impacts of tourism development on local communities is that of price inflation. The increased demand that results from tourists visiting an area usually leads to an increase in prices for goods and services. This can impose economic hardship on host communities, particularly if they do not receive greater incomes as a result of the tourism.

The term 'leakage' has come to be used to describe the loss of economic welfare from a host community when the money spent on visiting that community goes elsewhere. This is extremely common in marine tourism situations, because visitors often arrive in boats that they have provisioned with supplies elsewhere. In addition, in many island areas, marine tourism operators are not locals but are seasonal businesses that have their base elsewhere. Consequently much of the money spent on the business actually flows outside the host community.

The lack of spending in local communities has resulted in much controversy with regard to the cruise-ship industry. These large vessels are self-contained and thus, while large numbers of tourists may go ashore to visit a small community, little money is spent on food, accommodation and transport.

One further issue is that the economic benefits that arise from tourism development are seldom distributed evenly throughout communities and regions. There are, however, a number of locations that have derived significant economic benefits from the development of marine tourism activities. Kaikoura is such a location.

Case study: Kaikoura, New Zealand

Kaikoura is a small coastal community located around a small peninsula on the northeastern coast of New Zealand's South Island. The continental shelf is close to shore near this peninsula, coming as close

Plate 5.3 Tourists line up for one of four daily whale-watch trips at Kaikoura.

as a kilometre from shore just south of the township. The rapid increase in depth associated with the shelf, from as little as 30 metres to over 1,000 in a short distance, and the convergence of offshore currents in the vicinity of Kaikoura produce an upwelling of nutrient-rich waters that supports a rich marine food chain. As a result of this abundant ecosystem, a number of species of whales, dolphins and seals can be found there relatively close to shore.

Commercial whale and dolphin watching began in 1988 at Kaikoura and the industry has grown quickly to become the most important economic activity in the area (Baxter, 1993). Before marine-mammal-based tourism was set up in the town, Kaikoura was economically depressed. The impact of the tourism growth on the town has been widespread:

> The impact of Whale Watch Kaikoura on the local township has been major. Prior to the establishment of the company, Kaikoura was seen as an economically depressed area. Businesses were on a downturn, and people were having to leave the area to get work. Following the establishment of the company which brought increasing numbers of tourists, businesses across the board began to benefit through increased sales ... A host of

> new businesses have sprung up in the accommodation and catering areas. New craft shops and takeaway bars, as well as novelty shops, have also appeared. Real estate prices have also increased, and the tempo of life has picked up a little. Without Whale Watch Kaikoura, I think our town may have eventually given up the ghost and died.
>
> (Hawke, 1995: 39)

Socio-economic indicators back up Hawke's impressions. In 1991, soon after the cetacean-based tourism began in Kaikoura, unemployment in the area was still higher than the national average for New Zealand. Household income was significantly lower than the New Zealand average; only 29 per cent of the district's residents reported incomes higher than $NZ30,000 per year, as opposed to 44 per cent nationally. Annual visitor numbers to Kaikoura in 1989 were estimated to be 10,000. These increased dramatically to over 100,000 per year by 1993. This influx of tourists has provided a significant economic boom to the district, with unemployment dropping significantly and household incomes rising dramatically. Kaikoura is no longer viewed as an economically depressed region, but as one of the 'boom towns' in New Zealand's South Island Warren and Taylor, 1994).

SUMMARY

So, whilst there is an acceptance that marine and other nature-based tourism is growing, there is still considerable debate about, first, whether this kind of tourism is even desirable and, second, how it should be controlled to minimise detrimental impacts. For despite volumes of publications on tourism management in the past decade, 'we still know relatively little about how to control and manage tourism' (Butler, 1993: 43). As a result, a number of leading authors in the field are arguing for research in the nature-based tourism area. For example, Boo states that: 'despite rising expectations regarding the value of nature tourism in many fields of expertise, there are great gaps in the information necessary to manage the nature tourism industry' (Boo, 1990: 4). The debate, therefore, centres upon the type of tourism that should be created and permitted. A variety of management strategies can be employed in an attempt to negate detrimental impacts and to maximise the benefits of tourism. These approaches are outlined in the next chapter.

REVIEW QUESTIONS

1. Outline and discuss three negative and three positive impacts which can result from marine tourism.
2. Discuss the influence of 'perspective' with regard to the assessment of tourism impacts (for example, a tourism industry employee versus a local retiree).
3. Is the developmental path for tourism as outlined by Butler and others inevitable, or can locations develop differently? If so, what are the influential factors which render the development different? If not, why not?

6

MANAGEMENT APPROACHES

INTRODUCTION

Up to this point this book has demonstrated that marine tourism is large and growing rapidly, and furthermore that this kind of tourism is causing many problems. It has also been pointed out that not all impacts resulting from marine tourism are detrimental – in some cases the impacts are beneficial. The question arises, therefore, of what it is that causes the impacts of marine tourism to be positive or negative. The answer is – it depends. Sometimes the positive impacts are much greater than the negative simply because the tourism enterprise is very small, the resource large and consequently the use of the marine resource negligible. The benefits to the local community from the tourism may, however, be relatively large. Therefore, it may be concluded that the marine tourism operation is, on balance, beneficial. This judgement does, of course, depend on your perspective. If you are a local person deriving income from, say, a new, small-scale charter fishing operation, inevitably your perspective is inclined to the positive. If, however, you are a person who has retired to the location for peace, quiet and relaxation, your perspective on the benefits of local charter fishing business bringing more tourists to the area may be quite different. And, of course, if you take the extreme example, and put on not rose-coloured but marine-blue glasses and view things as a local reef fish, that perspective may view the small-scale fishing operation extremely negatively – especially if you are the fish that gets caught!

This simplistic illustration shows that judging the relative costs and benefits of tourism is extremely difficult, if not impossible. It is, however, possible to agree that a goal of maximising the positive and minimising the negative impacts of tourism is a worthy one. Thus, rather than trying to quantify costs and benefits it is better to concentrate on developing a management regime for marine tourism which maximises the good and minimises the bad. There is a wide variety of strategies that can be utilised in developing such a regime. This chapter details some of them. It also discusses the concepts of 'sustainability' and 'ecotourism', two terms that have become widespread in the laws and literature. These approaches seek to solve the tourism development dilemma of how to stop people destroying the very features which attract them to a location. This chapter also details a case for education as a management approach for marine tourism. Finally, some suggestions for measuring the impacts of tourism are made, and a conceptual model for its management is offered.

STRATEGIES USED TO MANAGE MARINE TOURISM

A number of well-established management regimes are utilised to manage tourists and their activities on the sea. These strategies range from no control whatsoever, to complex combinations of structures, technology, economics, regulation and education. In order to simplify this situation, Orams (1995) divided tourist management strategies into four main categories: regulatory, physical, economic and educational. The first two control tourist behaviour through external manipulation and have dominated most approaches to managing marine tourism in the past (for example, see Wallace, 1993). More recently, economic strategies have been utilised (Plimmer, 1992). The fourth category has traditionally been incorporated in park management activities (Beckmann, 1989) but has had limited application elsewhere.

Table 6.1 outlines these four categories and details a number of examples of specific management actions that can be grouped under these headings. The following sections discuss each category in more detail.

Regulatory management strategies

Regulatory management practices are the traditional method of controlling tourist activities in marine settings (Plimmer, 1992). Rules and regulations are commonly utilised to restrict visitor actions, access, times and numbers, and are usually posted on signs, notices and written material. Enforcement is usually undertaken by police, park rangers or other agents of the management authority. A system of progressively harsher punishment is most common; for example, a warning at the first offence, banning from the area at a second offence, and eventually fines and imprisonment in extreme cases of misconduct, such as harassment or killing of protected wildlife. Examples of regulations commonly used to manage tourists are visitor number restrictions, limits on times and locations, types of visitor activity permitted, spatial and temporal zoning, restrictions on the types of equipment permitted, noise levels, speed limits, permits, lease and licence requirements, and codes of practice.

The purpose of regulatory practices is threefold: first, to protect the safety of the tourist; second, to reduce conflicts between tourists; and

Table 6.1 Techniques for managing marine tourism

Technique	*Description*	*Examples*	*Purpose*	*Implication*
Physical				
Site hardening	Increasing the durability of the resource	Boardwalks, concrete launching ramps, mooring buoys	To reduce negative environmental impacts which result from intensive use (e.g. erosion)	Escalates development of site
Facility placement	Geographical location of facilities to 'channel' use to desired areas	Siting of boat ramps, marinas and moorings away from sensitive areas	To discourage use in sensitive/inappropriate areas	Concentrates use in specific areas
Facility design	Designing items to be safer, more durable or less harmful	Replacing children's old, wooden beach-play equipment with modern, plastic, purpose-designed equipment	To meet recreation needs better, improve safety and reduce vandalism/ deterioration	Usually more costly for recreation organisation initially – maintenance costs lower
Sacrifice areas	Allocating specific areas for intensive use in order that other sites remain pristine	Using above techniques to encourage majority of users to a specific beach, thereby reducing pressure on other nearby beaches	To 'sacrifice' a specific site by concentrating use so that other areas have lower levels of use	Rapid deterioration of sacrifice site Common spill-over to neighbouring sites Increased possibility of conflicts between users
Remove/alter attraction	Reducing the motivation for people to visit a particular site by changing the attraction	Moving 'Christ of the Deep' statue from coral reef to sand bottom in Pennekamp Marine Park, Florida	To decrease high-density use of specific sites	Public outcry re. removal of attraction; potential harm to attraction

Table 6.1 continued

Technique	*Description*	*Examples*	*Purpose*	*Implication*
Rehabilitation	Actively renovating an area to improve its quality and mitigate deterioration	Revegetation/planting programmes; reintroduction of endangered species	To improve quality of resource	Active programme needed Site often can not be used during rehabilitation
Regulatory				
Limit visitor numbers	Setting maximum levels for a site and close it to additional use after the limit has been reached	All mooring sites having to be booked in advance – once they are booked no more are permitted	To control impacts by restricting numbers of people	Does not cater for demand Casual users tend to miss out
Prohibit certain activities	Banning activities which may be harmful or unsafe or impact detrimentally on others	Banning use of all motorised water craft within 100 m of shore	To reduce conflict/harm	Resentment from groups which are banned Enforcement needed
Close areas to activities/use	Closing a specific area for all use or for certain uses for a specified time	Fencing off a dune area for six months to allow vegetation to regenerate	To allow areas to recover Reduce impacts from certain activities	Enforcement needed Displaces harmful use to other locations
Separate activities	Geographically or temporarily separating activities	Zoning areas for specific kinds of use; allowing swimming on certain days, windsurfing on others	To separate incompatible uses	Education and enforcement needed Reduces freedom of choice

Require minimum skill level	Restricting use of an area to people with a certain training/certification skill level	Visitors having to be certified in water safety/survival Visitors having to hold scuba certification or CPR/First Aid	To ensure skills match challenges To reduce negative impacts	Appropriate training courses must be available
Economic				
Differential fees	Charging higher fees for certain groups, activities, times or locations	A discounted boat-ramp fee for off-peak use	To spread use To ensure costs of managing certain activities are paid for by participants	'User pays' philosophy not widely accepted in some countries Reduces access for lower socio-economic groups
Damage bond	Requiring a damage deposit which is refunded to user if site is left in desired state	A $100 beach-use deposit which is refunded to the user only if inspection reveals site is left in suitable state	To provide a financial incentive for good behaviour To provide money for cleaning/rehabilitation if needed	Financial system and inspection service required
Fines	Imposing financial penalty for inappropriate/damaging behaviour	A fine for littering, vandalism or another regulation transgression	To penalise harmful acts	Enforcement needed Legislative backing needed
Rewards	Offering a financial reward for reporting inappropriate behaviour or undertaking desired behaviour	A prize for the greatest amount of litter collected by a group during a week A reward for reporting vandalism	To assist with enforcement of regulations An incentive for good behaviour	Finance needed

Table 6.1 continued

Technique	*Description*	*Examples*	*Purpose*	*Implication*
Education				
Printed material	Distributing printed material to visitors which describes/encourages appropriate behaviour	Brochures handed to all visitors which prompt them to take rubbish home	To encourage appropriate behaviour to reduce visitor impacts/conflicts	Need access to visitors before and during visit
Low-power radio	Broadcasting important information to visitors via AM radio band	Messages about weather, pollution or recent problems in area	To encourage appropriate behaviour to reduce visitor impacts/conflicts	Need access to visitors before and during visit
Signs	Displaying printed messages in appropriate locations	'Dunes being rehabilitated – please stay on track'	To encourage appropriate behaviour to reduce visitor impacts/conflicts	Important that wording is positive and sign does not detract from experience
Visitor centres	Structure which forms focal point for area's education efforts	Marine Park Visitor Centre	To encourage appropriate behaviour to reduce visitor impacts/conflicts	Major financial cost
Guided walks/talks	Formal communication programme from staff to visitors	Guided walk to seal colony	To encourage appropriate behaviour to reduce visitor impacts/conflicts	High quality of person's teaching skills imperative
Activities	Any activity designed to entertain and educate	Instruction in surf life-saving techniques	To encourage appropriate behaviour to reduce visitor impacts/conflicts	High quality of person's teaching skills imperative
Personal contact	General contact and communication between staff and visitors	Answering questions about the best beaches/ reefs to visit	To encourage appropriate behaviour to reduce visitor impacts/conflicts	Availability of staff critical

third, to protect the marine environment from negative impacts due to inappropriate tourist behaviour. As a result of the increasing use of marine areas for recreation and the increasing demand for opportunities for commercial tourist operations, there has been a proliferation of regulations pertaining to marine activities. These regulations often restrict the freedom of visitors to do as they wish and there is some evidence that this may reduce enjoyment of the experience (Hatten and Hatten, 1988). Other difficulties can arise from the cost of enforcing rules, particularly in large and remote areas such as those that exist in marine locations. Nevertheless, regulation remains the most common management response to increasing pressure from tourism (McArthur and Hall, 1993).

Physical management strategies

'Physical' approaches to management are those human-made structures that control human activity by restricting the movement or type of activity which can be undertaken. A typical marine example is the construction of a boardwalk across a wetland. This physical structure directs and facilitates the movement of tourists and reduces the negative impacts that could be caused by their walking through sensitive areas. Additional examples include underwater observatories, mooring buoys for vessels in coral reef areas, sea pens, tethered animals (sea turtles) and grandstands, such as those provided at the Phillip Island Penguin Colony in Victoria, Australia (personal observation).

Because of the difficulties involved with erecting structures in the marine environment, physical controls are not a particularly common system in controlling marine tourism. They are, however, very commonly used in terrestrial protected natural area management (Yale, 1991; Buckley and Pannell, 1992; Burgess, 1992; McArthur and Hall, 1993). In a number of situations physical structures have been successfully used to control tourists in the sea. The use of mooring buoys to reduce anchor damage on coral reefs is a notable success story (Salm and Clark, 1989). Additional examples include the use of such things as glass-bottom boats, semi-submersibles, self-guided underwater trails and beach bicycle pathways.

Human-made structures are also utilised to provide additional opportunities and services for tourists. Examples are marinas, wharves, boat ramps and observation platforms. Of particular relevance are vessels which tourists use to access the marine environment. Regulations can be used to restrict the type of vessel permitted in an area; for example, by allowing only electrically powered boats in an area sensitive to noise disturbance. Thus, the negative impacts of an activity can be mitigated by a combination of regulatory and physical approaches.

Economic management strategies

Economic strategies use prices as incentives or disincentives to modify people's behaviour. Although these techniques have received little explicit recognition, they have been used in many natural areas for many years (Plimmer, 1992). An example of this type of strategy is the use of higher entry fees to facilities during peak use times in an attempt to spread visiting. Permits that are auctioned to commercial tourist operators can restrict the number of operators. Imposing fines for littering, taking undersize fish, or other inappropriate behaviour is another example of a regulation combined with an economic disincentive. Although the use of economic strategies has not been common in marine tourism situations, considerable potential exists. For example, discounts on access fees to a marine park can be provided if groups undertake a clean-up project, or assist with research, during their visit. Fees could be more expensive during times when wildlife is more sensitive to disturbance, thus providing an incentive for tourists to visit at other times. Most marine resources are managed by publicly funded government organisations. Given the increasing financial pressure under which many of these public management agencies find themselves, the opportunity to utilise economic techniques to generate additional funds, and accomplish management objectives, may be worthwhile.

Educational management strategies

The goals of education-based management strategies are to reduce the incidence of inappropriate tourist behaviour by encouraging a voluntary behaviour change, and to increase visitor enjoyment and understanding. Many authors, therefore, view education as a potential 'win–win' situation for both the marine environment and the tourist (Forestell, 1990). However, the use of education as a management strategy in tourism situations has not been as common as the use of physical or regulatory techniques. This is due to a number of factors that make the planning and implementation of an effective educational programme for tourists particularly difficult (Beckmann, 1988). These factors include the diverse characteristics of tourist groups such as different group size, age and educational attainment. As a result, the needs of each tourist are unique and are difficult to cater for in the design of an educational programme. This is further complicated by the 'non-captive' nature of marine tourists; that is, tourists are usually free to come and go as they please and will often 'vote with their feet' and leave when educational programmes do not meet their needs or hold their interest. In marine settings, the diverse locations and the mobility and geographical spread of tourists often make it difficult for an education programme to be scheduled at a time, or located at a place, where tourists will be exposed to it.

Lack of knowledge regarding the marine environment and a lack of suitably trained and qualified personnel can also complicate efforts to implement an effective education programme (Beckmann, 1989). These factors result in little use of education as a management strategy and the predominance of physical and regulatory approaches. However, a number of authors argue that education is, or should be, a critical component of nature-based tourism experiences (Roggenbuck, 1987; O'Laughlin, 1989; Alcock, 1991; Oliver, 1992; Bramwell and Lane, 1993). This argument is further developed later in this chapter, where the role of 'interpretation' is discussed.

Case study: the reefs of the Florida Keys

The coral reefs of the Florida Keys are the only extensive coral reef tract in North America and the third largest reef system in the world. They contain many unique and endangered species, host intensive recreational use and support a significant commercial tourism industry producing an estimated $US1 billion per year for the local area (Belleville, 1991). The Keys' marine environment attracts more than three million visitors each year (Laycock, 1991) and there is widespread concern over the impact this high level of tourist use is having on the Keys' ecosystem. Ward's comments typify the reaction of many to the masses of tourists visiting the reefs:

> Their boats pollute the water and everything in it with petroleum products and sewage. Incompetent operators crash into the reefs. They litter the sea with plastic foam cups, aluminum cans, glass, plastic bags, bottles, and miles of tangled fishing line. This debris does not go away – it is, for all practical purposes, indestructible. Thousands of swimmers routinely bump, scrape, and step on coral.
> (Ward, 1990: 123)

In 1990, as a result of an Act of the United States Congress, the Florida Keys were declared a national marine sanctuary. National marine sanctuaries in the US are administered by the National Oceanic and Atmospheric Administration, an agency of the federal government. The central management tool for marine sanctuaries is a comprehensive management plan. The plan establishes the specific objectives, policies and guidelines for the sanctuary and helps to determine the regulations to be enforced. These management plans, therefore, consider the range of management strategies outlined earlier in this chapter and adopt those which are considered most appropriate for protecting the resources of the area whilst allowing for traditional commercial activities, such as tourism, to continue.

The management plan for the Florida Keys National Marine Sanctuary (National Oceanic and Atmospheric Administration, 1995) outlines a wide variety of strategies to manage tourist use of the area. Fundamental to these strategies is the establishment of a system of five types of geographic zone. This regulatory approach designates specific areas for specific uses. *Wildlife management areas* have restricted access and activities in order to protect endangered or threatened species and their habitats. *Replenishment reserves* protect spawning and nursery areas by minimising human influences and restricting extractive practices such as fishing. *Sanctuary preservation areas* protect areas for non-extractive recreational use. *Special use areas* designate areas for specific purposes such as scientific research. *Existing management areas* allow for activities to continue under current management restrictions.

Additional 'action plans' set out in the management plan include channel marking (physical approach), a mooring buoy action plan (physical approach), a regulatory action plan (regulatory approach), an enforcement action plan (regulatory and economic approaches), an education action plan (educational approach) and ones dealing with water quality, research and monitoring, submerged cultural resources and volunteers, each of which contains elements of all four management approaches.

The Florida Keys is a large area that is intensively used for tourism. A comprehensive management plan that sets out a variety of management strategies is a common technique used in marine resource management.

MARINE PARKS

One of the most popular and successful management regimes utilised both to protect resources and to facilitate recreational use of them is the establishment of marine parks. These locations are administered in many different forms. One extreme is the designation of a 'no-take' marine protected area where no disturbance or removal of marine life is permitted, such as the Cape Rodney to Okakari Point Marine Reserve at Leigh, New Zealand (Ballantine, 1991). The other extreme is illustrated by the adjacent Hauraki Gulf Maritime Park, offshore from Auckland, which provides no special protection or management of marine resources but seeks to manage the publicly owned islands of the area (Hauraki Gulf Maritime Park Board, 1983).

Within these extremes there is a wide variety of institutional arrangements for managing marine resources and marine recreation (Sorensen and McCreary, 1990). The most popular is the 'multi-use' marine protected area (Salm and Clark, 1989). These marine parks seek to manage recreational and commercial use of the marine resources whilst protecting them from unacceptable damage. The most popular mechanism for doing this is establishing a range of geographic zones (see the case study above for an example in the Florida Keys).

This kind of management approach has received widespread support at both a scientific and a political level. In January 1992, then United States President George Bush declared that 'in the United States, marine sanctuaries and other protected areas offer one of the best methods of safeguarding the marine environment' (Orams, 1993: 4). Marine parks are, therefore, now seen by many as an important framework under which various management techniques can be applied to ensure that use of the marine environment, including tourist activities, is sustainable. The largest marine park in the world is the Great Barrier Reef Marine Park.

Case study: the Great Barrier Reef Marine Park

The Great Barrier Reef (GBR) is the world's largest coral reef system, stretching 2,000 km down the northeastern coast of Queensland, Australia. The reef is actually comprised of some 2,600 individual reefs that vary greatly in size (Kelleher, 1987). It supports a diverse range of marine species, including around 1,500 fish species, 350 species of hard corals and 240 species of birds (Kelleher, 1987). Estimates in the early 1990s placed the number of visitors to the reef at 2.2 million per year. By the year 2000 this number is expected to have doubled (Dinesen, 1995).

In 1975 the Australian government established the Great Barrier Reef Marine Park (GBRMP). The park is managed by the Great Barrier Reef Marine Park Authority (GBRMPA). However, the GBRMPA shares responsibility for the management of the park area with the state of Queensland (Dinesen, 1995). The broad management tools used to control tourism and recreation in the park include zoning plans, management plans, regulations, special area designations, permits and education (Dinesen, 1995).

Zoning plans require considerable public consultation and establish permitted uses along each section of the reef. For example, some zones exclude permanent structures. Zoning plans also establish which tourist activities require permits.

Management plans are prepared for individual reefs or islands that are likely to receive significant use for tourism. Historically, the management of tourism impacts has been largely controlled with the use of permits for commercial tourist operators because management plans have been slow to develop.

In the early years of the park the relationship between use of the park's resources and impacts was not well understood. Permits were therefore favoured as a means of managing impacts because they allowed flexibility. Over time the number of permits being issued and their complexity increased dramatically. The process of permit granting and renewal was also being slowed down with appeals. A further concern was the inequity of the process. Commercial tourism operators were being prevented from offering activities in some areas where private individuals could undertake the same activities (Dinesen, 1995).

In high-use areas of the GBR, the GBRMPA has adopted a number of physical methods to reduce impacts. In 1989, four permanent large-vessel moorings and six small-boat moorings were erected in close proximity to dive sites. These were designed to prevent coral being damaged by boat anchors (Alder and Haste, 1995). Fishing in the area was also restricted in order to protect the area's resident populations of potato cod. In 1992, it was estimated that this site, an area around 300 metres in length, received 30,000 dives (Alder and Haste, 1995).

A further management initiative for the GBR was taken by the commercial tourism operators themselves. In 1992, they grouped together to form a reef operators' association. With the help of the GBRMPA they instituted a degree of self-regulation. For example, they placed restrictions on boat sizes, on length of stay at particular reefs and on fish feeding. This technique has proved an effective way of reducing the negative impacts associated with commercial tourist use of the reef (Alder and Haste, 1995).

Educational approaches have also been emphasised by the GBRMPA. A number of studies have shown that these techniques have been effective; for example, research showed that educating scuba divers reduces the impact that they have on coral reefs (Medio, Ormond and Pearson, 1997; Rouphael and Inglis, 1995).

The GBR is a large resource with rapidly growing use and associated impacts. However, through the establishment of a marine park it has an agency with a mandate to manage human use of that resource comprehensively. This organisational structure and the use of a wide variety of management techniques are seen as an important model of how tourism can be managed sustainably in marine environments.

MARINE TOURISM AND SUSTAINABILITY

The objective of sustainability in managing natural resources has become widespread in recent years (World Tourism Organisation, 1997c). Derived from the influential Brundtland Report (United Nations World Commission on Environment and Development, 1987) and ratified by the Rio de Janeiro Environment Summit in 1992 (UNCED, 1992), the concept of sustainability has become a guiding principle in legislation, policies and approaches to a wide range of resource management topics (World Tourism Organisation, 1997c). Sustainability is defined as 'a characteristic of a process or state that can be maintained indefinitely' (IUCN, UNEP and WWF, 1991: 211). Sustainable use is defined as 'use of an organism, ecosystem or other renewable resource at a rate within its capacity for renewal' (IUCN, UNEP and WWF, 1991: 211). Tourism has also been considered in terms of the principle of sustainability (World Tourism Organisation, 1997c; Stabler, 1997) and has been viewed as a potentially sustainable industry (Burnie, 1994). However, a number of authors question the ability of any tourism venture to be sustainable. For example, Zell states: 'Tourism creates more tourism, the location becomes well known and thus desirable creating demand, more supply and ultimately destruction of the original reason for going there' (Zell, 1992: 31).

Partly in response to the push for 'sustainability' in resource use and management, and partly as a result of the recognition that tourism can result in significant environmental impacts that are not 'sustainable' in the longer term, the concept of 'ecotourism' has arisen.

Ecotourism

Ecotourism has been hailed by some as the 'answer' to tourism development, and 'its supporters argue that ecotourism is the only tourism development that is sustainable in the long term' (Warren and Taylor, 1994: 1). The concept that tourism should contribute to the health and viability of the natural attraction upon which it is based is an appealing one. However, these lofty aspirations and the rapid adoption of this label by many operators and nations have been greeted with cynicism by many, who view ecotourism as simply nature-based tourism 'dressed up' under a new, attractive label:

> Ecotourism is big business. It can provide foreign exchange and economic reward for the preservation of natural systems and wildlife. But ecotourism also threatens to destroy the resources on which it depends. Tour boats dump garbage in the waters off Antarctica, shutterbugs harass wildlife in National Parks, hordes of us trample fragile areas. This frenzied activity threatens the viability of natural systems. At times we seem to be loving nature to death.
>
> (Berle, 1990: 6)

Berle's statement typifies the concerns of many regarding the increasing number of tourists who are visiting natural areas and who are having a detrimental impact on those areas. Others, such as Wight (1993), caution that the ecotourism label is being utilised to take advantage of a 'greening' of the economic market place and to 'eco-sell' tourism and travel. In some cases ecotourism may well be nothing more than a new marketing gimmick which dresses up existing tourism attractions in an attempt to increase market share:

> There is no question that 'green' sells. Almost any terms prefixed with the term 'eco' will increase interest and sales. Thus, in the last few years there has been a proliferation of advertisements in the travel field with references such as ecotour, ecotravel, ecovacation, ecologically sensitive adventures, eco(ad)ventures, ecocruise, ecosafari, ecoexpedition and, of course, ecotourism.
>
> (Wight, 1993: 4)

Johnston agrees and states that:

> In some cases people have been blindly applying the buzzwords ecotourism, soft-path tourism and sustainable development to national and international development schemes, promoting and acquiring funding for these schemes because of the labels they wear.
>
> (Johnston, 1990b: 4)

Although many countries and agencies look towards ecotourism as an answer to both economic and conservation objectives (Boo, 1990), many remain unconvinced that such ventures are a panacea that both protects the environment and supports economic activity (Butler, 1990; Zell, 1992; Pearce, 1994; Wheeller, 1994). Considerable debate exists, therefore, over whether ecotourism can be sustainable, and over what management regimes and strategies can be employed to minimise the negative impacts associated with anthropogenic influences on natural ecosystems.

As a result of the rapidly increasing use of resources and the desire to protect them, a number of leading authors in the field are arguing for research in the ecotourism area. For example, Boo states that: 'despite rising expectations regarding the value of nature tourism in many fields of expertise, there are great gaps in the information necessary to manage the nature tourism industry' (Boo, 1990: 4). The debate therefore centres upon the type of tourism that ecotourism is, or should be. A variety of management strategies, such as those outlined above, can be employed in an attempt to negate detrimental impacts. However, it is possible that educational approaches are best suited to the management of nature-based tourism.

The case for education as a management strategy

Educational approaches to management have long been utilised in the park management field. In fact, the profession has coined the term 'interpretation' to describe the use of educational approaches to natural resource management.

Interpretation is a word traditionally used to describe the process of translating meaning from one spoken language into another. Interpretation of the natural environment in a park management context is a similar kind of communication. Although the activity of interpretation has existed since pre-biblical times (Weaver, 1982), the first explicit use of the term and discussion of its meaning were made by Freeman Tilden, who defined interpretation as: 'An educational activity which aims to reveal meanings and relationships through the use of original objects, by firsthand experience, and by illustrative media, rather than simply to communicate factual information' (Tilden, 1957: 9). He suggests, therefore, that interpretation is a particular type of education that focuses on meanings and relationships. It is not necessarily about the communication of facts, but about the communication and learning of ideas and concepts and the imparting of an appreciation for the natural environment involved.

What is not provided by Tilden's definition is an explanation of the practical management objectives of interpretation. Interpretation has, in fact, multiple roles which go beyond imparting an appreciation for nature (McArthur and Hall, 1993). In particular, it can assist in achieving management objectives. For example, Beckmann, who conducted a review of interpretation activity in Australian parks, states:

> Most Australian interpreters believe that interpretation has a real role to play in the management of parks (e.g. by reducing the need for regulation and enforcement, increasing visitor awareness of appropriate behaviour, enabling careful distribution of visitor pressure to minimize environmental impacts on fragile natural resources).
> (Beckmann, 1989: 148)

Despite widespread support of interpretation as a management strategy for natural areas (reported examples include Pope, 1981; Price, 1985; Whately, 1987; Beckmann, 1988, 1989; Jelinek, 1990; Burgess, 1992), there has been little empirical research which has demonstrated the specific benefits of interpretation programmes (Uzzell, 1989). 'Unfortunately, very few applications of interpretation to management in Australia are fully documented, and the full potential of interpretation as "the public face of management" has still to be realized' (Beckmann, 1989: 148). This is most likely to be related to the difficulties in undertaking research in natural area settings and in determining such things as the incidence of visitor behaviour that is counter to agency management objectives, visitor enjoyment and visitor knowledge, attitudes and understanding.

There is, therefore, a case for assessing the effectiveness of education-based management strategies in managing marine tourism. The objectives of such strategies are sound and the potential exists to protect the marine environment, increase visitor enjoyment and understanding, and prompt more environmentally responsible behaviour.

Creating more responsible marine tourists

At a basic level, the overall goal of management strategies designed to control marine tourism is twofold: first, to protect the marine environment from detrimental impacts, and second, to provide for and promote enjoyable tourist experiences. However, in order to accomplish this ambition tourists must somehow be prompted to do more than simply 'have a good time'. Furthermore, and equally important, commercial operators must see their responsibilities as extending beyond making a profit. They must see their role as one of contributing to the health and viability of the resource upon which the livelihood of their business depends. If this can be achieved, the environment, upon which the tourism is based, would actually benefit from the tourism.

This view is contentious and certainly would be viewed with cynical contempt or, at the very least, as overly optimistic by some (for example, see Wheeller, 1992, 1994). However, in certain specific circumstances, nature is already benefiting from tourism. For example, the Wildfowl Trust in Gloucestershire, England, has successfully assisted in the recovery of a number of endangered bird species, and now reintroduces many back into the wild to supplement threatened wild populations (Yale, 1991). These kinds of programme, largely funded by tourists, are becoming more common. In addition, a number of habitats, which might have otherwise been lost, have been set aside as wildlife parks because of their tourism value, thereby benefiting the wildlife (for example, see Groom, Podolsky and Munn, 1991; Barnes, Burgess and Pearce, 1992).

The fact is that many tourists are willing to give of their time, money and labour in order to assist with nature conservation. It may not, therefore, be overly optimistic to aim for marine-based tourism that is mutually beneficial. Given suitable management strategies, it may be possible. The success of a management regime can, therefore, be measured in terms of its effectiveness in prompting the movement of the tourist experience towards these desired objectives. The first of these is changing tourists' behaviour and lifestyle so that their actions become more environmentally responsible, both during the tourism experience and, longer-term, after the experience. The second is promoting tourist actions that directly contribute to the environment while they are visiting it; for example, through

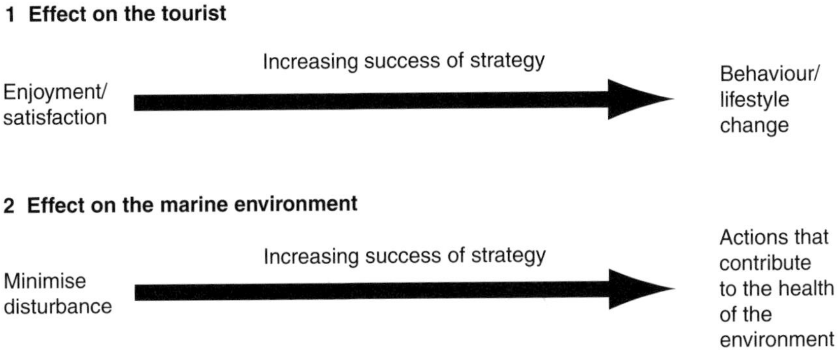

1 Effect on the tourist

Enjoyment/ satisfaction → Increasing success of strategy → Behaviour/ lifestyle change

2 Effect on the marine environment

Minimise disturbance → Increasing success of strategy → Actions that contribute to the health of the environment

Figure 6.1 Objectives of marine tourism management strategies (Orams, 1995)

assisting with research projects, becoming involved with habitat restoration programmes, removing litter, acting as volunteers for the management agency, or even helping to police the area so that other tourists' actions are not detrimental to the environment. These desired objectives are illustrated in Figure 6.1.

The categorisation of current management practices persented earlier in this chapter (see Table 6.1) showed that a number of authors place significant emphasis upon educational techniques as a suitable mechanism for managing human impacts on the natural environment. This optimism has seldom been empirically tested (Uzzell, 1989). The view that an effective interpretation programme for tourists will result in better tourism is, however, commonly held. The potential exists to prompt more desirable tourism through a management strategy that is based upon educational techniques.

One of the few pieces of research which has attempted to assess the effectiveness of an education-based management regime on marine tourism was conducted at Tangalooma, Queensland, Australia.

Case study: the dolphins of Tangalooma

Tangalooma is a tourist resort in southeastern Queensland, Australia. Since 1992 a group of wild bottlenose dolphins have been regular visitors to the beach adjacent to this resort (Orams, 1994). The dolphins visit the area nightly to receive fish handouts from tourists in shallow water beside the resort's pier. This opportunity has been promoted as an attraction and it has become increasingly popular since its inception.

The dolphin feeding at Tangalooma was used as an opportunity to assess the effectiveness of an environmental education programme as a mechanism to promote environmentally desirable changes in tourists' attitudes and behaviour. The study compared tourists who were exposed to the education programme (experiment group) with those who were not (control group). Indicators of changes in levels of enjoyment, knowledge, attitudes, intentions and behaviour were measured and compared between control and experiment groups in order to assess the impact of the education programme.

The study showed that interacting with dolphins produced a desire in tourists to change their behaviour and become more environmentally responsible. However, those tourists who were not given the structured education programme seldom carried out these good intentions. In contrast, many of the tourists who were given the education programme became significantly more 'green' in their behaviour, by doing things such as joining environmental groups and getting involved in environmental issues (Orams, 1997).

This study suggests that marine tourism experiences without a structured education component are unlikely to produce the changes in tourists' behaviour proposed in Figure 6.1 above. However, the study does show that a structured education programme can prompt tourists to become more environmentally responsible and move the tourism experience to the more desired state identified in Figure 6.1.

Plate 6.1 The nightly feeding of the dolphins at Tangalooma has become a popular tourist attraction.

MEASURING THE IMPACTS OF MANAGEMENT

In order to measure the success of management strategies in achieving a shift towards the objectives for tourism illustrated in Figure 6.1, a number of indicators need to be selected, for, as Pearce states:

> Tourism impacts have been widely studied, but comparatively few attempts have been made to incorporate these impacts into tourism typologies. If this were more commonly done, it would facilitate bringing together cause and effect ... In each case, specific measures of each variable used must be established, either in quantitative or qualitative terms.
>
> (Pearce, 1994: 25)

These specific measures or 'outcome indicators' are shown in Figure 6.2. This figure shows the transition to the desirable form of ecotourism as a series of steps. As a first step, a management regime can be measured in terms of its impact on tourist satisfaction and enjoyment. However, this should not be viewed as the primary objective of such a strategy. More complex educational and behavioural objectives should be given equal weighting. The intermediate steps which assist in a transition from mere enjoyment to actual behaviour change that benefits the environment are, initially, the facilitation of education and learning, and subsequently, the changing of attitudes and beliefs to those that are more environmentally and ecologically sound. These four steps (or indicators) – satisfaction/enjoyment, education/learning, attitude/belief change and behaviour/lifestyle change – are categories under which research instruments can be designed to measure the effectiveness of a management strategy in achieving the transition illustrated in Figure 6.2.

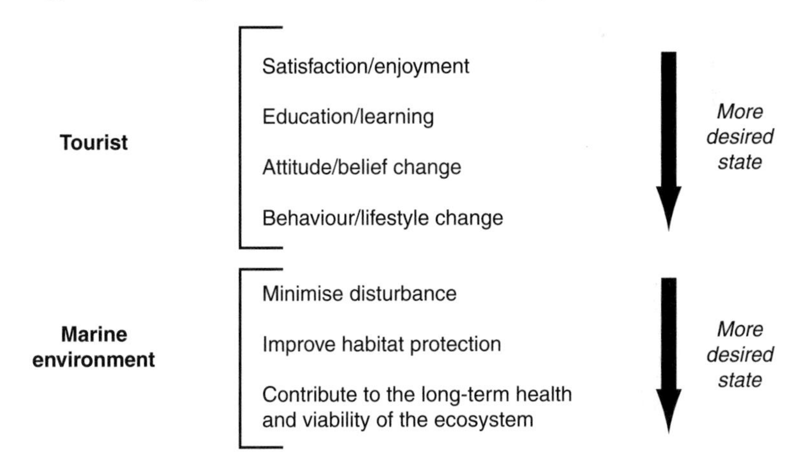

Figure 6.2 Outcome indicators (Orams, 1995)

The research instruments used to measure the achievement of these outcomes could include such things as participant observation, questionnaires and interviews. Of particular importance in assessing behaviour change are follow-up surveys which assess the impact of the experience some time after the tourists have returned home. For this purpose, a follow-up interview or questionnaire needs to be conducted. It is well recognised that intentions to change behaviour do not necessarily result in actual behaviour change (Gudgion and Thomas, 1991). The follow-up research should attempt to gather information on whether the desired behaviour change actually occurred.

The second set of steps shown in Figure 6.2 is that of those that assist in measuring the impact of the tourists' behaviour on the marine environment. The objective is to encourage a transition from a passive position, where the tourist simply seeks to minimise detrimental impacts on the environment, to an active one, where the tourist actually contributes to the health and viability of the environment. These three categories – minimising detrimental impacts, improving habitat quality and comprehensively contributing to the long-term health and viability of the ecosystem – are far more difficult to design measurement instruments for. Indices will need to be drawn up on a case-by-case basis; however, management regimes which are judged to fulfil the last criterion should be viewed as more desirable than those that merely seek to minimise detrimental impacts.

Each environment upon which tourism is based will be different and, therefore, indicators selected to monitor progress towards the desired state will need to be carefully chosen. However, for each setting, decisions should be made on, first, what types and levels of change in the natural ecosystem are acceptable; second, what critical indicators should be used to monitor this change; and third, what human actions are appropriate and inappropriate for that setting. Various techniques can be used to make these decisions. For example, if adequate scientific knowledge is available for the resource, the most important indicators may already be known. In reality, however, this is seldom the case.

Processes such as those developed for environmental and strategic planning (including public participation) can be used, or an 'expert' panel can be developed to arrive at a consensus on the indicators. In this way hypotheses can be developed which allow testing of the level of disturbance to the natural environment.

Assessing whether tourists' actions improve habitat quality/protection is less difficult. If tourists provide financial support and/or labour which directly assists in the maintenance, protection or improvement of the natural resource they are visiting, they are meeting this objective. The more desirable state occurs when a comprehensive and on-going programme exists where tourists can not only contribute during their visit but continue to support and contribute finance, labour and expertise to the resources on a long-term basis. Measurement of these indicators involves assessing the management regimes in place for specific tourism programmes to establish whether they facilitate this kind of tourist involvement.

A CONCEPTUAL MODEL FOR THE MANAGEMENT OF MARINE TOURISM

This chapter has attempted to review the range of strategies available for managing marine tourism. It has also proposed that management strategies should attempt to do more than simply minimise negative impacts. It argues that marine tourism can be managed to be, on balance, a positive thing for the marine environment. It is recognised that this is extremely ambitious and perhaps unrealistic given the rapid growth and intensity of use demonstrated in earlier chapters. However, the alternative, namely to accept that our enjoyment of the seas will result in their eventual destruction, is less palatable. Educational strategies in combination with other approaches show considerable potential for creating the sustainable 'ecotourism' that so many hope is attainable.

When considering the management of marine tourism activities this statement of Plimmer's is pertinent:

> So, we have a wide range of management techniques. We can add to them as we realise the possibilities. It is essential that we look at all these possible techniques as a menu, and choose the one, or combination, best suited to the situation.
>
> (Plimmer, 1992: 125)

This pragmatic view is backed by the comments of Ceballos-Lascurain (1993), who argues that managing tourism requires a multidisciplinary approach. This chapter attempts to provide a brief review of some of the approaches that may be a 'menu' from which specific strategies can be designed.

Earlier, in Chapter 4, the Spectrum of Marine Recreation Opportunities (Table 4.1) was outlined. This model, combined with ones of management techniques (Table 6.1) and outcome indicators (Figure 6.2), provides a conceptual framework for considering the management of marine tourism (Figure 6.3). This framework is useful in clarifying the marine recreational activity being discussed (the Spectrum of Marine Recreation Opportunities), the human intervention being undertaken (the management options available) and the outcomes that are being sought (outcome indicators). This model represents a conceptual framework that can be used as a basis for understanding and assessing the effectiveness of differing management regimes.

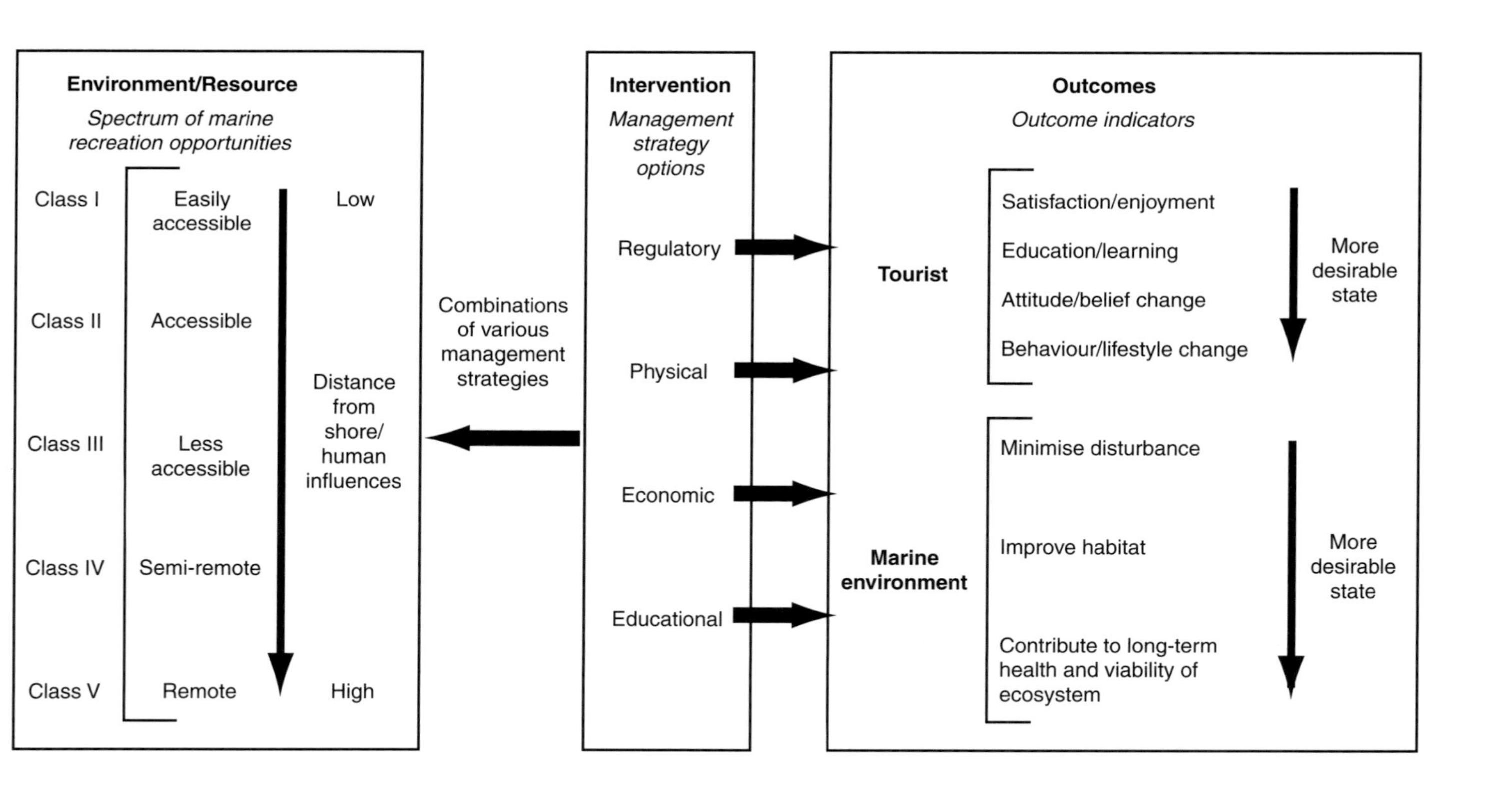

Figure 6.3 A conceptual model for the management of marine tourism

SUMMARY

The growth of marine tourism is a world-wide phenomenon that is likely to continue into the twenty-first century, becoming one of the major economic and social forces of coastal nations. Problems associated with it are widely acknowledged, and the concept of ecotourism has arisen, in part, as a result of a desire to create tourism enterprises which are based upon the natural environment but which do not degrade or destroy it.

A wide variety of opportunities for tourists to recreate in the sea exists, and the demand is considerable. The issue of how these kinds of activity should be managed remains unanswered. An argument that marine tourists should be encouraged, through the management strategies in place in a marine area, to become active contributors to the health and viability of the marine environment has been made. It is possible that marine interpretation programmes are the best mechanism to achieve this.

REVIEW QUESTIONS

1 Name the four categories of management strategy and give an example for each.
2 Discuss the concept of ecotourism. Is it useful in improving the management of marine tourism activities?
3 What is the potential of education in managing marine tourism? Do you see any special challenges in implementing such strategies in marine settings?
4 Discuss the importance of monitoring the effectiveness of management strategies.

7
THE FUTURE

PROBLEMS AND CHALLENGES

Despite the significant efforts being made to mitigate the impacts of marine tourism in programmes such as coastal zone management, marine park and protected areas, management plans and fisheries management systems, the reality of ever-increasing numbers and fixed resources remains. Our interest in marine recreation seems unlikely to abate. Increasing interest in and use of the marine environment has been a long-term trend. Furthermore, it is certain that our invention of new ways to access the sea and utilise it for recreational purposes will continue to increase. The result of these two basic trends is simple: marine-based tourism will continue to grow in popularity. Because of the increasing demand, the supply of marine tourism opportunities will become a critical issue.

Plate 7.1 Opportunities for marine tourism experiences in high-quality natural environments may become scarce in the future.

Supply of marine tourism opportunities is constrained by several important factors. First, marine resources are limited. Whilst the oceans are vast, the locations sought for marine recreation are relatively small. A significant factor in this will be environmental quality. Marine tourists do not want to conduct their recreational activities in polluted areas, because many of the attractions they seek are not present in polluted locations and their own health is threatened there. Thus, management of marine resources in order to maintain or improve environmental quality will become the major challenge.

A further issue with regard to marine tourism will be cost. A basic economic principle – high demand and scarce supply – will continue to force the cost of marine recreational activities up. Many cultures regard free access to and use of marine resources as a basic human right. However, it is already obvious that access to high-quality areas and popular activities, in some areas, is only afforded by the wealthy.

A further issue derived from increasing demand and limited supply is the conflict between incompatible activities. This is common in popular areas now, and will escalate over the coming decades. Underwater photographers who value large fish species alive will compete with spear fishers who wish to hunt them. Jet-skiers who wish to wave ride and wave jump will endanger surfers who wish to use the same 'resource'. Wildlife watchers will conflict with water-skiers who wish to use sheltered bays for their sport. Indeed, all of these conflicts are common today. They will be an increasing challenge for marine resource managers in the future.

An additional factor that should be noted is that most degradation of marine resources is not the result of tourists nor their activities. The damage caused by the pollution of our coastal environs from human activities on land and from commercial use of our oceans for fishing, the dumping of waste, dredging and so on far outweighs the influence of tourism. Consequently, the future of marine tourism is inextricably linked with all other human activities that affect the sea. To give a crude example, the whale-watching industry cannot exist if commercial whaling practices wipe out all the whales. One of the most significant challenges, and perhaps one of the greatest opportunities, for marine tourism is to ensure that the demand for the use of marine resources for recreational purposes is given due consideration in decision making on marine resource utilisation.

While the future for marine resources looks bleak, there are a number of cases which show positive results. As has occurred on land, the uniting of large numbers of people who wish to use natural environments for recreation and who value high-quality natural environments has produced positive change for nature. The marine conservation movement, while not as well established as terrestrial equivalents, appears to be gathering momentum. 'Save the oceans' campaigns, beach clean-ups, marine mammal and sea bird rehabilitation centres and many other efforts have shown that humans who are interested in things marine can make a positive difference through their efforts.

One of the most encouraging signs of positive change is the increasing number of marine parks and marine protected areas that are being established worldwide. When the decisions of humans over the past two hundred years are reviewed, close to the top of the list of 'good decisions' will be the establishment and protection of large areas of high natural quality as national parks and other protected areas. These not only provide a habitat for many of the planet's species, but also significantly contribute to the health and functioning of 'Gaia' – and of all things that live as a part of it.

As our use of marine resources for tourism and for other needs continues to grow, marine protected areas will also become critical to the health of 'Gaia'. They will provide locations that are not only important for marine recreational activities but, most significantly and as their terrestrial counterparts have done, make it possible to nurture a conservation ethic in those who visit them. The role of marine resource managers, environmental groups, management agencies and marine tourism operators is critical. If we are to exist as a species for the next thousand years we must utilise every opportunity to change the way people behave towards things marine. If people are not inspired and changed as a result of their marine tourism experiences then the industry is simply exploitative and will ultimately become destructive.

The year 1998 was declared by the United Nations as the International Year of the Oceans. This is timely. As we approach the next millennium, the use of our oceans needs to be carefully considered. Tourism has an important role to play in the use of those resources. The industry could become an agent for positive change – a contributor to the healthy functioning of 'Gaia' – or it could be yet another significant cause of the decline of the quality of life on our planet. Given the size and diversity of the industry, it is likely to be both. The way that it is managed will make the difference as to which role predominates.

POTENTIAL SOLUTIONS AND AREAS FOR RESEARCH

A number of management solutions have been posed earlier (see Chapter 6). In particular, the potential of education-based management strategies has been emphasised. However, it would be naive to presume that well-educated marine tourists will solve all problems associated with the industry. Complex challenges seldom have simple solutions, and consequently the marine tourism dilemma of conserving marine attractions whilst allowing for their use will never be solved completely. Rather, the problem is more like a battle where resource managers must continually work to minimise negative impacts and maximise positive impacts, by developing strategies from the range of techniques available to meet the unique challenges provided by individual situations.

In fighting that battle, answers to a number of important questions will provide additional weapons for resource managers. These questions include the following:

- What motivates people to pursue marine recreation?
- What are the characteristics of marine tourists?
- What techniques are most effective in controlling human behaviour in marine settings?
- How can the impacts of recreational activities on marine resources be minimised?
- How can the negative impacts of tourism development on host communities be minimised?
- How can the benefits of marine tourism activities be maximised?
- What techniques are most effective in reducing conflict between competing uses?
- What management regimes or decision-making approaches are most effective?

The role of marine scientists is crucial in providing answers to these questions and thus adding to the 'menu' of management strategies available to marine resource managers. It is ironic that while almost all of the challenges faced by the marine environment are the result of human activities, including tourism, the great majority of research that occurs on our oceans remains in the biological and physical sciences. It is critical for our marine environment that marine scientists incorporate and encourage the social sciences in their community. Increasing our understanding of humans, what they do and why, is fundamental to finding answers to the challenges posed by marine tourism and other human uses of marine resources. In particular, a focus on applied research – that which is focused on providing solutions to problems rather than simply outlining the extent of the problems – is paramount.

SUMMARY

This book has provided a brief overview of a very broad topic. It has adopted a conservation-oriented perspective to the issues surrounding marine tourism because I consider the health of our seas a prerequisite to the survival of the industry. The challenges faced in managing marine tourism are huge, and there will probably always be more tales of failure than cases of success. However, because there are success stories, where the management of the tourism activity has resulted in positive change for local communities and for marine ecosystems, there is hope. If it can be done in one area, perhaps it can be done in others. Continued investigation, critical thinking, learning from experiences and communicating those experiences are ways forward to a better future.

REVIEW QUESTIONS

1 Discuss three of the challenges outlined for marine tourism. Do you agree or disagree that they are important?
2 Are there any additional research questions that will be important for the future management of marine tourism?
3 What do you think are the most effective ways of managing marine tourism? Discuss your approach to the challenges outlined in this chapter.

BIBLIOGRAPHY

Adler, J. 1989. 'Origins of sightseeing'. *Annals of Tourism Research* 16: 7–29.

Albert, D.M. and Bowyer, R.T. 1991. 'Factors related to Grizzly bear–human interactions in Denali National Park'. *Wildlife Society Bulletin* 19: 339–349.

Alcock, D. 1991. 'Education and extension: management's best strategy'. *Australian Parks and Recreation* 27(1): 15–17.

Alder, J. and Haste, M. 1995. 'The Cod Hole: a case study in adaptive management'. In O. Bellwood, H. Choat and N. Sascena (eds), *Recent Advances in Marine Science and Technology '94*. PACON International, Honolulu, Hawaii, and James Cook University of North Queensland, Townsville.

Amante-Helweg, V.L.U. 1995. 'Cultural perspectives of dolphins by ecotourists participating in a "swim with wild dolphins" programme in the Bay of Islands, New Zealand'. MA thesis, University of Auckland.

Anderson, J. 1878. *Zoological Results of the Two Expeditions to Western Yunnan in 1868 and 1875*. Quaritch, London.

Anderson, J. 1994. *Participation in Watersports*. Insights, English Tourist Board, London.

Anonymous 1991. 'The SCUBA of the future'. *Geographical Magazine* October: 10.

Archer, B.H. 1989. 'Tourism and small island economies'. In C.P. Cooper (ed.), *Progress in Tourism, Recreation and Hospitality Management. Vol. 1*. Belhaven Press, London.

Bairstow, J. 1986. 'Packing more punch'. *High Technology*, August: 29–30.

Ballantine, W.J. 1991. *Marine Reserves for New Zealand*. University of Auckland Press, Auckland.

Barnes, J., Burgess, J. and Pearce, D. 1992. 'Wildlife tourism'. In T.M. Swanson and E.B. Barbier (eds), *Economics for the Wilds: Wildlife, Wildlands, Diversity and Development*. Earthscan Publications, London.

Baxter, A.S. 1993. 'The management of whale and dolphin watching at Kaikoura, New Zealand'. In D. Postle and M. Simmons (eds), *Encounters with Whales '93*. Great Barrier Reef Marine Park Authority, Townsville.

Beckmann, E.A. 1988. 'Interpretation in Australia: some examples outside national parks'. *Australian Parks and Recreation* 24(3): 8–12.

Beckmann, E.A. 1989. 'Interpretation in Australian national parks and reserves: status, evaluation and prospects'. In D.L. Uzzell (ed.), *Heritage Interpretation. Vol. 1: The Natural and Built Environment*. Belhaven Press, London.

Belleville, B. 1991. 'Too beautiful to live'. *Sunshine Magazine of South Florida*. 16 June: 12–16.

Berle, P.A.A. 1990. 'Two faces of ecotourism'. *Audubon* 92(2): 6.

Boo, E. 1990. *Ecotourism: The Potential and Pitfalls*. World Wide Fund for Nature, Washington DC.

Borge, L., Nelson, W.C., Leitch, J.A. and Leistritz, F.L. 1991. *Economic Impact of Wildlife Based Tourism in Northern Botswana*. Agricultural Economics Report No. 262. Agricultural Experiment Station, North Dakota University, North Dakota.

Bramwell, B. and Lane, B. 1993. 'Interpretation and sustainable tourism: the potential and the pitfalls'. *Journal of Sustainable Tourism* 1(2): 71–80.

Brasch, R. 1995. *How Did Sports Begin?*. HarperCollins, Sydney.

Brent Wheeler and Co. 1997. *The Economic Impact of the America's Cup on the Auckland Region and Territorial Local Authorities Within the Region*. Report to Auckland Regional Services Trust. Brent Wheeler and Co. Ltd, Auckland.

Buckley, R. and Pannell, J. 1992. 'Environmental impacts of tourism and recreation in national parks and conservation reserves'. In B. Weiler (ed.), *Ecotourism Incorporating the Global Classroom*. International Conference Papers. Bureau of Tourism Research, Canberra.

Burger, J. and Gochfield, M. 1993. 'Tourism and short term behavioural responses of nesting masked, red-footed, and blue-footed boobies in the Galapagos'. *Environmental Conservation* 20(3): 255–259.

Burgess, J. 1992. 'Softly softly, minimising the impacts of tourism in Tasmania'. In B. Weiler (ed.), *Ecotourism Incorporating the Global Classroom*. International Conference Papers. Bureau of Tourism Research, Canberra.

Burgett, J. 1990. 'Diving deeper into Hanauma: identifying conceptual barriers to effective management'. In M. Miller and J. Auyong (eds), *Proceedings of the 1990 Congress on Coastal and Marine Tourism. Vol. 1*. National Coastal Resources Research Institute, Corvallis OR.

Burnie, D. 1994. 'Ecotourists to paradise'. *New Scientist* 16 April: 24–27.

Butler, R.W. 1980. 'The concept of a tourist area cycle of evolution'. *Canadian Geographer* 24(1): 5–12.

Butler, R.W. 1990. 'Alternative tourism: pious hope or Trojan horse?'. *Journal of Travel Research* 28(3): 40–45.

Butler, R.W. 1993. 'Tourism – an evolutionary perspective'. In J.G. Nelson, R.W. Butler and G. Wall (eds), *Tourism and Sustainable Development: Monitoring, Planning, Managing*. Heritage Resources Centre Joint Publication No. 1. University of Waterloo, Ontario.

Caribbean Tourism Organisation 1993. *Caribbean Tourism Statistical Report 1992*. CTO, St Michael, Barbados.

Carvallo, M.L. 1994. 'Antarctic tourism must be managed, not eliminated'. *Forum for Applied Research and Public Policy* Spring: 76–79.

Ceballos-Lascurain, H. 1993. 'Overview of ecotourism around the world: IUCN's ecotourism program'. In *Proceedings of the 1993 World Congress on Adventure Travel and Ecotourism*. Adventure Travel Society, Englewood CO.

Cherfus, J. 1984. *Zoo 2000*. British Broadcasting Corporation, London.

Clamen, E. and Rossier, E. 1991. 'Animal-tourism and leisure in rural areas'. Equ'idee – *Bulletin d'Information sur les Equides* 2: 42–49.

Connor, R.C. and Smolker, R.S. 1985. 'Habituated dolphins (*Tursiops* sp.) in Western Australia'. *Journal of Mammalogy* 66(2): 398–400.

Corbin, A. 1994. *The Lure of the Sea*. Polity Press, Cambridge.

Cousteau, J.Y. 1985. *The Ocean World*. Harry Abrams, New York.

Crandall, R., Nolan, M. and Morgan, L. 1980. 'Leisure and social interaction'. In S.E. Iso-Ahola (ed.), *Social Perspectives on Leisure and Recreation*. Charles Thomas, Springfield IL.

Croall, J. 1995. *Preserve or Destroy: Tourism and the Environment*. Calouste Gulbenkan Foundation, London.

Crompton, J. 1979. 'Motivations for pleasure vacations'. *Annals of Tourism Research* 6: 408–424.

Csikszentmihalyi, M. 1975. *Beyond Boredom and Anxiety*. Jossey-Bass, San Francisco.

Cuthbert, A. and Suman, D. 1995. 'To jet ski or not to jet ski: personal watercraft conflicts in the lower Florida Keys'. In D. Suman, M. Shirlani and M. Villanueva (eds), *Urban Growth and Sustainable Habitats: Case Studies of Policy Conflicts in South Florida's Coastal Environment*. Rosenstiel School of Marine and Atmospheric Science, University of Miami, Miami.

Davies, M. 1990. 'Wildlife as a tourism attraction'. *Environments* 20(3): 74–77.

Davis, D. and Tisdell, C. 1995. 'Recreational scuba-diving and carrying capacity in MPAs'. *Ocean and Coastal Management* 26(1): 19–40.

Deci, E. 1975. *Intrinsic Motivation*. Plenum, New York.

Dignam, D. 1990. 'Scuba gaining among mainstream travellers'. *Tour and Travel News* 26 March.

Dinesen, Z. 1995. 'New approaches to managing tourism impacts in the Great Barrier Reef Marine Park'. In O. Bellwood, H. Choat and N. Sascena (eds), *Recent Advances in Marine Science and Technology '94*. PACON International, Honolulu, Hawaii, and James Cook University of North Queensland, Townsville.

Doak, W. 1984. *Encounters with Whales and Dolphins*. Hodder and Stoughton, Auckland.

Dowling, R. 1992. 'An ecotourism planning model'. In B. Weiler (ed.), *Ecotourism Incorporating the Global Classroom*. International Conference Papers. Bureau of Tourism Research, Canberra.

Doxey, G.V. 1975. 'A causation theory of visitor–resident irritants'. In *Proceedings of the Travel and Tourism Research Association Sixth Annual Conference*. Travel and Tourism Research Association, San Diego.

Duffus, D.A. and Dearden, P. 1990. 'Non-consumptive wildlife oriented recreation: a conceptual framework'. *Biological Conservation* 53: 213–232.

Duffus, D.A. and Dearden, P. 1993. 'Recreational valuation, and management of killer whales (*Orcinus orca*) on Canada's Pacific coast'. *Environmental Conservation* 20(2): 149–156.

Duffus, D.A. and Wipond, K.J. 1992. 'A review of the institutionalization of wildlife viewing in B.C. Canada'. *Northwestern Environmental Journal* 8(2): 325–345.

Duffy, E. 1957. 'The psychological significance of the concept of arousal or activation'. *Psychological Review* 64: 265–275.

Dyer, P.K. 1992. 'Wedgetailed shearwater nesting patterns: a spatial and ecological perspective, Capricorn Group, Great Barrier Reef'. Ph.D. thesis. University of Queensland.

Earle, S.A. 1995. 'Sea change'. *Ocean Realm*, June: 30–34.

Ellmers, D. 1991. 'The Deutsches Schiffahrts Museum, Bremerhaven'. In P. Neill and B.E. Krohn (eds), *Great Maritime Museums of the World*. Balsam Press, New York.

Ewert, A.W. 1989. *Outdoor Adventure Pursuits: Foundations, Models and Theories*. Publishing Horizons, Columbus OH.

Fabbri, P. (ed.) 1990. *Recreational Uses of Coastal Areas*. Kluwer Academic, Dordrecht.

Forestell, P.H. 1990. 'Marine education and ocean tourism: replacing parasitism with symbiosis'. In M.L. Miller and J. Auyong (eds), *Proceedings of the 1990 Congress on Coastal and Marine Tourism. Vol. 1*. National Coastal Resources Research Institute, Corvallis OR.

Forestell, P.H. and Kaufman, G.D. 1990. 'The history of whalewatching in Hawaii and its role in enhancing visitor appreciation for endangered species'. In M.L. Miller and J. Auyong (eds), *Proceedings of the 1990 Congress on Coastal and Marine Tourism. Vol. 1*. National Coastal Resources Research Institute, Corvallis OR.

Forestell, P.H. and Kaufman, G.D. 1995. 'Whale watching in Hawaii as a model for development of the industry worldwide'. In K. Colgan, S. Prasser and A. Jeffery (eds), *Encounters with Whales '95*. Australian Nature Conservation Agency, Canberra.

Frigden, J. and Hinkelman, B. 1977. 'Recreation behaviour and environment congruence'. Paper presented at the National Recreation and Parks Association Research Symposium, National Recreation and Parks Association, Las Vegas NV.

Gabbay, R. 1986. *Tourism in the Indian Ocean Island States of Mauritius, Seychelles, Maldives and Comoros*. University of Western Australia and National Centre for Development Studies, Islands Australia Project, Perth.

Gauthier, D.A. 1993. 'Sustainable development, tourism and wildlife'. In J.G. Nelson, R.W. Butler and G. Wall (eds), *Tourism and Sustainable Development: Monitoring, Planning and Managing*. Heritage Resources Centre Joint Publication No. 1. University of Waterloo, Ontario.

Gibson, J.D. 1976. 'Seabird islands no. 38, big island, five islands, New South Wales'. *Australian Bird Bander* 14: 100–103.

Gilbert, E.W. 1953. *Brighton: Old Ocean's Bauble*. Methuen, London.

Glasson, J., Godfrey, K. and Goodey, B. 1995. *Toward Visitor Impact Management*. Ashgate Publishing, Aldershot.

Gourlay, M.R. 1991. 'Coastal observations for monitoring environmental conditions on a coral reef island'. In *Proceedings of the Tenth Australasian Conference on Coastal and Ocean Engineering*. University of Auckland, Auckland.

Gray, H.P. 1970. *International Travel – International Trade*. Heath Lexington Books, Lexington ME.

Great Barrier Reef Committee 1977. *Conservation and Use of the Capricorn and Bunker Groups of Islands and Coral Reefs*. GBRC, Brisbane.

Griffin, R. 1992. 'Threatened coastlines'. *CQ Researcher*, 2(5): 97–120.

Griffiths, M. and Van Schaik, C.P. 1993. 'The impact of human traffic on the abundance and activity periods of Sumatran rain forest wildlife'. *Conservation Biology* 7(3): 623–626.

Groom, M.J., Podolsky, R.O. and Munn, C.A. 1991. 'Tourism as a sustained use of wildlife: a case study of Madre de Dios, southeastern Peru'. In *Neotropical Wildlife Use and Conservation*. University of Chicago Press, Chicago.

Gudgion, T.J. and Thomas, M.P. 1991. 'Changing environmentally relevant behaviour'. *Environmental Education and Information* 10(2): 101–112.

Guthier, D.A. 1993. 'Sustainable development, tourism and wildlife'. In J.G. Nelson, R.W. Butler and G. Wall (eds), *Tourism and Sustainable Development: Monitoring, Planning, Managing*. Heritage Resources Centre Joint Publication No. 1. University of Waterloo, Ontario.

Haddock, C. 1993. *Managing Risks in Outdoor Activities*. New Zealand Mountain Safety Council, Wellington.

Hall, C.M. and Johnston, M.E. (eds) 1995. *Polar Tourism: Tourism in the Arctic and Antarctic Regions*. John Wiley and Sons, Chichester.

Hammit, W.E., Dulin, J.N. and Wells, G.R. 1993. 'Determinants of quality wildlife viewing in Great Smokey Mountains National Park'. *Wildlife Society Bulletin* 21(1): 21–30.

Hanna, N. and Wells, S. 1992. 'Sea sickness'. *Focus (Tourism Concern)* 5: 4–6.

Hatten, K.J. and Hatten, M.L. 1988. *Effective Strategic Management: Analysis and Action*. Prentice Hall, Englewood Cliffs NJ.

Hauraki Gulf Maritime Park Board 1983. *The Story of Hauraki Gulf Maritime Park*. Hauraki Gulf Maritime Park Board, Auckland.

Hawke, L. 1995. 'Whale watch Kaikoura'. In K. Colgan, S. Prasser and A. Jeffery (eds), *Encounters with Whales '95*. Australian Nature Conservation Agency, Canberra.

Heath, R.A. 1992. 'Wildlife-based tourism in Zimbabwe: an outline of its development and future policy options'. *Geographical Journal of Zimbabwe* 23: 59–78.

Heatwole, H. and Walker, T.A. 1989. Dispersal of alien plants to coral cays. *Ecology* 70: 787–790.

Hegerl, E.J. 1984. 'An evaluation of the Great Barrier Reef Marine Park concept'. In W.T. Ward and P. Saenger (eds), *The Capricornia Section of the Great Barrier Reef Marine Park: Past, Present and Future*. Royal Society of Queensland and Coral Reef Society, Brisbane, Queensland.

Hendee, J. and Roggenbuck, J. 1984. 'Wilderness related education as a factor increasing demand for wilderness'. Paper presented at the International Forest Congress Convention, Quebec City, Canada, 5 August.

Heron Island Resort 1986. Advertising brochure.

Hopley, D. 1988. 'Anthropogenic influences on Australia's Great Barrier Reef'. *Australian Geographer* 19: 26–45.

Houston, J.R. 1996. 'The economic value of U.S. beaches'. In J. Auyong (ed.), *Abstracts of the 1996 World Congress on Coastal and Marine Tourism*. Oregon Sea Grant, Oregon State University, Corvallis OR.

Hoyt, E. 1996. 'Whale watching: a global overview of the industry's rapid growth and some recent implications and suggestions for Australia'. In K. Colgan, S. Prasser and A. Jeffery (eds), *Encounters with Whales '95*. Australian Nature Conservation Agency, Canberra.

Hulsman, K. 1983. 'Survey of seabird colonies in the Capricornia section of the Great Barrier Reef Marine Park II: population parameters and management strategies'. Unpublished research report to the Great Barrier Reef Marine Park Authority, Townsville, Queensland.

Hulsman, K. 1984. 'Seabirds of the Capricornia section of the Great Barrier Reef Marine Park'. In W.T. Ward and P. Saenger (eds), *The Capricornia Section of the Great Barrier Reef Marine Park: Past, Present and Future*. Royal Society of Queensland and Australian Coral Reef Society, Brisbane, Queensland.

Hunt, J. McV. 1969. *The Challenge of Incompetence and Poverty*. University of Illinois Press, Urbana IL.

Ingold, P., Huber, B., Neuhaus, P., Mainini, B., Marbacher, H., Schnidrig-Petrig, R. and Zeller, R. 1993. 'Tourism and sport in the alps – a serious problem for wildlife?'. *Revue Suisse de Zoologie* 100(3): 529–545.

Iso-Ahola, S. 1982. 'Toward a social psychological theory of tourism motivation: a rejoinder'. *Annals of Tourism Research* 12: 256–262.
Iso-Ahola, S. 1989. 'Motivation for leisure'. In E.L. Jackson and T.L. Burton (eds), *Understanding Leisure and Recreation: Mapping the Past, Charting the Future*. Venture Publishing, State College PA.
IUCN, UNEP and WWF (World Conservation Union, United Nations Environment Programme and World Wide Fund for Nature) 1991. *Caring for the Earth: A Strategy for Sustainable Living*. Gland.
Jeffery, A. 1993. 'Beyond the breach – managing for whale conservation and whale watching in Hervey Bay Marine Park, Qld'. In D. Postle and M. Simmons (eds), *Encounters with Whales '93*. Workshop Series No. 20. Great Barrier Reef Marine Park Authority, Townsville, Queensland.
Jelinek, A. 1990. 'An interpretation emphasis for park management'. *Australian Parks and Recreation* 26(4): 32–33.
Jenner, P. and Smith, C. 1992. *The Tourism Industry and the Environment*. Special Report No. 2453. Economist Intelligence Unit, London.
Johnston, B.R. 1990a. '"Save our beach dem and our land too!" The problems of tourism in "America's Paradise"'. *Cultural Survival Quarterly* 14(1): 6–10.
Johnston, B.R. 1990b. 'Introduction: breaking out of the tourist trap'. *Cultural Survival Quarterly* 14(1): 2–5.
Jolliffe, I.P., Patman, C.R. and Smith, A.J. (eds) 1985. *The Coastal Zone: Challenge, Management and Change*. Scottish Academic Press, Edinburgh.
Jones, B.L. 1993. 'The emerging undersea leisure industry'. *Sea Technology* February: 38–42.
Jones, O.A. 1967. 'The Great Barrier Reef Committee – its work and achievements, 1922–1926'. *Australian Natural History* 15: 315–318.
Kavanagh, T. 1997. 'Dollars roll in as surf champs pull tourists'. *Sunday Mail* 22 June: 44.
Kelleher, G. 1987. 'Management of the Great Barrier Reef Marine Park'. *Australian Parks and Recreation* 23(5): 27–33.
Kelleher, G. 1990. 'Floating hotels on the Great Barrier Reef'. In S.D. Halsey and R.B. Abel (eds), *Proceedings of the International Symposium on Coastal Ocean Space Utilization*. Elsevier, New York.
Kenchington, R.A. 1990a. 'Tourism in coastal and marine environments: a recreational perspective'. In M.L. Miller and J. Auyong (eds), *Proceedings of the 1990 Congress on Coastal and Marine Tourism. Vol. 1*. National Coastal Resources Research Institute, Corvallis OR.
Kenchington, R.A. 1990b. *Managing Marine Environments*. Taylor and Francis, New York.
Kenchington, R.A. 1991. 'Tourism development in the Great Barrier Reef Marine Park'. *Ocean and Shoreline Management* 15: 57–78.
Kerr, L. 1991. 'Ducks don't vote: the dilemma of wildland wildlife managers'. *Trends* 28(2): 30–34.
Keys, N. 1987. 'Shark Bay: looking beyond piecemeal planning'. *Habitat Australia* 15(5): 16–18.
Kim, S. and Kim, Y.J.E. 1996. 'Overview of coastal and marine tourism in Korea'. *Journal of Tourism Studies* 7(2): 46–53.
Lane, S.G. 1991. 'Some problems during the exodus of young shearwaters from Mutton Bird Island, New South Wales'. *Corella* 15: 108.

Laycock, G. 1991. 'Good times are killing the Keys'. *Audubon* 93(5): 38–41.

Leiper, N. 1995. *Tourism Management*. TAFE Publications, Collingwood, Victoria.

Leopold, A. 1949. *A Sand County Almanac*. Oxford University Press, Oxford.

Limpus, C.J., Fleay, A. and Guinea, M. 1984. 'Sea turtles of the Capricornia section, Great Barrier Reef Marine Park'. In W.T. Ward and P. Saenger (eds), *The Capricornia Section of the Great Barrier Reef Marine Park: Past, Present and Future*. Royal Society of Queensland and Australian Coral Reef Society, Brisbane.

Lockhart, D.G. and Drakakis-Smith, D. (eds) 1997. *Island Tourism: Trends and Prospects*. Pinter, London.

Lovelock, J.E. 1979. *Gaia: A New Look at Life on Earth*. Oxford University Press, Oxford.

Lovelock, J.E. 1985. 'Elements'. In N. Myers (ed.), *The Gaia Atlas of Planet Management*. Pan, London.

Lundberg, D.E. and Lundberg, C.B. 1993. *International Travel and Tourism*. John Wiley and Sons, New York.

McArthur, S. and Hall, C.M. 1993. 'Visitor management and interpretation at heritage sites'. In C.M. Hall and S. McArthur (eds), *Heritage Management in New Zealand and Australia: Visitor Management, Interpretation and Marketing*. Oxford University Press, Auckland.

McKegg, S., Probert, K., Baird, K. and Bell, J. 1996. 'Marine tourism in New Zealand: environmental issues and options'. In J. Auyong (ed.), *Abstracts of the 1996 World Congress on Coastal and Marine Tourism*. Oregon Sea Grant, Oregon State University, Corvallis OR.

Madsen, J.S. 1991. 'The Viking Ship Museum'. In P. Neill and B.E. Krohn (eds), *Great Maritime Museums of the World*. Balsam Press, New York.

Major, B. 1995. 'New ships are fully loaded and ready to set sail'. *Cruise Desk* November: 8–9.

Manning, R.E. 1986. *Studies in Outdoor Recreation*. Oregon State University Press, Corvallis OR.

Marti, B.E. 1992. 'Passenger perceptions of cruise itineraries'. *Marine Policy* September: 360–370.

Mathieson, A. and Wall, G. 1982. *Tourism: Economic, Physical and Social Impacts*. Longman, New York.

Mattix, R. and Goody, J.M. 1990. 'Conflicts in ocean recreational use in Kaneoke Bay: a community dilemma'. In M.L. Miller and J. Auyong (eds), *Proceedings of the 1990 Congress on Coastal and Marine Tourism*. National Coastal Resources Research and Development Institute, Newport OR.

Medio, D., Ormond, R.F.G. and Pearson, M. 1997. 'Effect of briefings on rates of damage to corals by SCUBA divers'. *Biological Conservation* 79(1): 91–95.

Mellor, B. 1990. 'Loving the reef to death?'. *Time* 45: 48–55.

Miles, J.C. 1990. 'Wilderness'. In J.C. Miles and S. Priest (eds), *Adventure Education*. Venture Publishing, State College, PA.

Miller, M.L. 1990. 'Tourism in the coastal zone: portents, problems, and possibilities'. In M.L. Miller and J. Auyong (eds), *Proceedings of the 1990 Congress on Coastal and Marine Tourism. Vol. 1*. National Coastal Resources Research Institute, Corvallis OR.

Miller, M.L. 1993. 'The rise of coastal and marine tourism'. *Ocean and Coastal Management* 20(3): 181–199.

Miller, M.L. and Auyong, J. (eds) 1990. *Proceedings of the 1990 Congress on Coastal and Marine Tourism.* National Coastal Resources Research Institute, Corvallis OR.

Miller, M.L. and Auyong, J. 1991. 'Coastal zone tourism: a potent force affecting environment and society'. *Marine Policy* March: 75–99.

Miller, M.L. and Auyong, J. (eds) 1998. *Proceedings of the 1996 World Congress on Coastal and Marine Tourism.* Washington Sea Grant Program, University of Washington School of Marine Affairs, and Oregon State Grant Program, Seattle WA.

Miller, M.L. and Kaae, B.C. 1993. 'Coastal and marine ecotourism: a formula for sustainable development?'. *Trends* 30(2): 35–41.

Morris, R. 1988. 'Human contact'. In R. Harrison and M.M. Bryden (eds), *Whales, Dolphins and Porpoises.* Facts on File, New York.

Muir, F. 1993. 'Managing tourism to a seabird nesting island'. *Tourism Management* 14(2): 99–105.

Myers, N. (ed.) 1985. *The Gaia Atlas of Planet Management.* Pan, London.

National Oceanic and Atmospheric Administration 1995. *Florida Keys National Marine Sanctuary. Draft Management Plan/Environmental Impacts Statement.* United States Department of Commerce, Washington DC.

Neil, D.T. 1988. 'Holothurian populations on Heron Reef, GBR: effect of the boat channel'. Unpublished report to the Australian Marine Sciences Association meeting, 28 May.

Neil, D.T. 1993. 'Barrier Reef studies'. Unpublished course handbook. Department of Geographical Sciences and Planning, University of Queensland, Brisbane, Queensland.

Neil, D.T., Orams, M.B. and Baglioni, A.J. 1995. 'Effect of previous whale watching experience on participants' knowledge of, and response to, whales and whale watching'. In K. Colgan, S. Prasser and A. Jeffery (eds), *Encounters with Whales '95.* Australian Nature Conservation Agency, Canberra.

Neill, P. 1991. 'The South Street Seaport Museum, New York'. In P. Neill and B.E. Krohn (eds), *Great Maritime Museums of the World.* Balsam Press, New York.

O'Halloran, T. 1996. 'Atlantis adventures: development of undersea tourism attractions'. In J. Auyong (ed.), *Abstracts of the 1996 World Congress on Coastal and Marine Tourism.* Oregon Sea Grant, Oregon State University, Corvallis OR.

O'Laughlin, T. 1989. 'Walk softly – but carry a big education campaign'. *Australian Ranger Bulletin* 5(3): 4–7.

Oliver, J. 1992. 'All things bright and beautiful: are tourists getting responsible adult environmental education campaigns?'. In B. Weiler (ed.), *Ecotourism Incorporating the Global Classroom.* International Conference Papers. Bureau of Tourism Research, Canberra.

Orams, M.B. 1993. 'Towards a marine conservation ethic: our marine protected areas can lead the way'. *Trends* 30(2): 4–7.

Orams, M.B. 1994. 'Tourism and marine wildlife: the wild dolphins of Tangalooma'. *Anthrozoos* 7(2): 195–201.

Orams, M.B. 1995. 'A conceptual model of tourist–wildlife interaction: the case for education as a management strategy'. *Australian Geographer* 27(1): 39–51.

Orams, M.B. 1997. 'The effectiveness of environmental education: can we turn tourists into "greenies"?'. *Progress in Tourism and Hospitality Research* 3: 295–306.

Orams, M.B. and Forestell, P.H. 1995. 'From whale harvesting to whale watching: Tangalooma 30 years on'. In O. Bellwood, H. Choat and N. Saena (eds), *Recent Advances in Marine Science and Technology '94.* PACON International, Honolulu, Hawaii, and James Cook University of North Queensland, Townsville.

O'Shea, T.J. 1995. 'Waterborne recreation and the Florida manatee'. In R.L. Knight and K.J. Gutzwiller (eds), *Wildlife and Recreationists*. Island Press, Washington DC.

Pattullo, P. 1996. *Last Resorts: The Cost of Tourism in the Caribbean*. Cassell, London.

Pearce, D.G. 1994. 'Alternative tourism: concepts, classifications and questions'. In V.L. Smith and W.R. Eadington (eds), *Tourism Alternatives: Potentials and Problems in the Development of Tourism*. University of Pennsylvania Press, Philadelphia PA.

Peisley, T. 1995. 'Transport: the cruise ship industry to the 21st century'. *EIU Travel and Tourism Analyst* 2: 4–25.

Pleumarom, A. 1993. 'What's wrong with mass ecotourism'. *Contours (Bangkok)* 6(3/4): 15–21.

Plimmer, W.N. 1992. 'Managing for growth: regulation versus the market'. In *Proceedings of the Conference on Ecotourism Business of the Pacific*. University of Auckland, Auckland.

Pope, L.V. 1981. 'Interpretation of urban natural areas'. *Trends* 18(4): 29–31.

Price, D. 1985. 'Understanding the Everglades'. *Trends* 22(4): 29–31.

Puhipau. 1994. 'Boycott paradise'. *Focus* Summer: 10.

Reeves, R.R. 1992. *Whale Responses to Anthropogenic Sounds: A Literature Review*. Department of Conservation Science and Research Series No. 47. Department of Conservation, Wellington.

Reynolds, E. 1990. 'Hanauma Bay baseline users survey'. In M. Miller and J. Auyong (eds), *Proceedings of the 1990 Congress on Coastal and Marine Tourism. Vol. 1*. National Coastal Resources Research Institute, Corvallis OR.

Rockel, M.L. and Kealy, M.J. 1991. 'The value of non-consumptive wildlife recreation in the United States'. *Land Economics* 67(4): 422–434.

Roggenbuck, J.W. 1987. 'Park interpretation as a visitor management strategy'. In *Proceedings of the Sixtieth Annual Conference of the Royal Australian Institute of Parks and Recreation*. Royal Australian Institute of Parks and Recreation, Canberra.

Rouphael, T. and Inglis, G. 1995. *The Effects of Qualified Recreational SCUBA Divers on Coral Reefs*. CRC Reef Research Technical Report. James Cook University of North Queensland, Townsville.

Salm, R.V. and Clark, J.R. (eds) 1989. *Marine and Coastal Protected Areas: A Guide for Planners and Managers*. International Union for the Conservation of Nature and Natural Resources. Gland.

Sathiendrakumar, R. and Tisdell, C.A. 1990. 'Marine areas as tourist attractions in the southern Indian Ocean'. In M.L. Miller and J. Auyong (eds), *Proceedings of the 1990 Congress on Coastal and Marine Tourism. Vol. 1*. National Coastal Resources Research Institute, Corvallis OR.

Scherl, L.M. 1987. 'Our need for wilderness: a psychological view'. *Habitat Australia* 15(4): 32–35.

Schiebler, S.A., Crofts, J.C. and Hollinger, R.C. 1996. 'Florida tourists' vulnerability to crime'. In A. Pizam and Y. Mansfield (eds), *Tourism Crime and International Security Issues*. John Wiley and Sons, Chichester.

Shackley, M. 1990. 'Manatees and tourism in south Florida: opportunity or threat?'. In M.L. Miller and J. Auyong (eds), *Proceedings of the 1990 Congress on Coastal and Marine Tourism. Vol. 2*. National Coastal Resources Research Institute, Corvallis OR.

Shackley, M. 1996. *Wildlife Tourism*. International Thomson Business Press, London.

Shaw, W.W. 1984. 'Problems in wildlife valuation in natural resource management'. In G.W. Peterson and A. Randall (eds), *Valuation of Wildland Resource Benefits*. Westview Press, Boulder CO.

Smith, C. and Jenner, P. 1994. 'Tourism and the environment'. *EIU Travel and Tourism Analyst* 5: 35–49.

Sorensen, J.C. and McCreary, S.T. 1990. *Institutional Arrangements for Managing Coastal Resources and Environments*. National Park Service, Washington DC.

Stabler, M.J. (ed.) 1997. *Tourism and Sustainability: Principles to Practice*. CAB International, Wallingford.

Stankey, G. 1985. *Carrying Capacity in Recreational Planning: An Alternative Approach*. United States Department of Agriculture – Forest Service, Ogden UT.

Tabata, R.S. 1990. 'Dive travel – a case study in educational tourism: policy implications for resource management and tourism development'. Paper presented at the Global Classroom Symposium, Christchurch, New Zealand, 19–22 August.

Thomas, L. 1985. Untitled. In N. Myers (ed.), *The Gaia Atlas of Planet Management*. Pan, London.

Thoreau, H.D. 1972 (reprint). *A Different Drummer*. April House.

Tilden, F. 1957. *Interpreting Our Heritage*. University of North Carolina Press, Chapel Hill NC.

Towner, J. 1996. *An Historical Geography of Recreation and Tourism in the Western World*. John Wiley and Sons, Chichester.

UNCED (United Nations Conference on Environment and Development) 1992. *Agenda 21*. UNCED, Rio de Janeiro.

United Nations World Commission on Environment and Development 1987. *Our Common Future: The Way Forward*. Oxford University Press, Oxford.

Uzzell, D.L. (ed.) 1989. *Heritage Interpretation. Vol. 2: The Visitor Experience*. Belhaven Press, London.

Vickerman, S. 1988. 'Stimulating tourism and economic growth by featuring new wildlife recreation opportunities'. *Transactions of the Fifty-Third American Wildlife and Natural Resources Conference*: 414–423.

Viskovic, N. 1993. 'Zootourism'. *Turizam* 41(1–2): 23–25.

Wallace, G.N. 1993. 'Visitor management: lessons from Galapagos National Park'. In K. Lindberg and D.E. Hawkins (eds), *Ecotourism: A Guide for Planners and Managers*. Ecotourism Society, Boulder CO.

Walton, J.K. 1983. *The English Seaside Resort: A Social History 1750–1914*. Leicester University Press, Leicester.

Walton, J.K. and Smith, J. 1995. 'The first century of beach tourism in Spain: San Sebastian and the Playas del Norte from the 1830s to the 1930s'. In M. Barke, J. Towner and M.T. Newton (eds), *Tourism in Spain: Critical Issues*. CAB International, Wallingford.

Ward, F. 1990. 'Florida's coral reefs are imperilled'. *National Geographic* 178(1): 115–132.

Warren, J.A.N. and Taylor, C.N. 1994. *Developing Ecotourism in New Zealand*. New Zealand Institute for Social Research and Development, Wellington.

Waters, S.R. 1990. *Travel Industry World Yearbook: The Big Picture – 1990. Vol. 34*. Child and Waters, New York.

Weaver, H.E. 1982. 'Origins of interpretation'. In G.W. Sharpe (ed.), *Interpreting the Environment*. John Wiley and Sons, New York.

West, N. 1990. 'Marine recreation in North America'. In P. Fabbri (ed.), *Recreational Uses of Coastal Areas*. Kluwer Academic, Amsterdam.

Whately, R. 1987. 'Coastal interpretation – some management implications'. *Australian Parks and Recreation* 23(4): 34–36.

Wheeller, B. 1991. 'Tourism's troubled times: responsible tourism is not the answer'. *Tourism Management* 12(1): 91–96.

Wheeller, B. 1992. 'Alternative tourism – a deceptive ploy'. In C.P. Cooper and A. Lockwood (eds), *Progress in Tourism, Recreation and Hospitality Management. Vol. 5*. Belhaven Press, London.

Wheeller, B. 1994. 'Ecotourism: a ruse by any other name'. In C.P. Cooper and A. Lockwood (eds), *Progress in Tourism, Recreation and Hospitality Management. Vol. 7*. Belhaven Press, London.

White, R. 1959. 'Motivation reconsidered: the concept of competence'. *Psychological Review* 56: 297–333.

Wight, P. 1993. 'Ecotourism: ethics or eco-sell?'. *Journal of Travel Research* 31(3): 3–9.

Wilks, J. 1993. 'Calculating diver numbers: Critical information for scuba safety and marketing programs'. *SPUMS Journal* 23(1): 11–14.

Woodland, D.J. and Hooper, J.N.A. 1977. 'The effect of human trampling on coral reefs'. *Biological Conservation* 11: 1–4.

World Tourism Organisation 1997a. *Yearbook of Tourism Statistics*. World Tourism Organisation, Madrid.

World Tourism Organisation 1997b. *Tourism: 2020 Vision*.World Tourism Organisation, Madrid.

World Tourism Organisation 1997c. *Agenda 21 for the Travel and Tourism Industry*. World Tourism Organisation/World Tourism Council/Earth Council, Madrid.

Wright, J.W. 1990. *The Universal Almanac*. Andrews and McMeel, New York.

Yale, P. 1991. *From Tourist Attractions to Heritage Tourism*. ELM Publications, Huntington.

Zell, L. 1992. 'Ecotourism of the future – the vicarious experience'. In B. Weiler (ed.), *Ecotourism Incorporating the Global Classroom*. International Conference Papers. Bureau of Tourism Research, Canberra.

Zuckerman, M. 1971. 'Dimensions of sensation-seeking'. *Journal of Consulting and Clinical Psychology* 36: 45–52.

Zuckerman, M. 1979. *Sensation-Seeking: Beyond the Optimal Level of Arousal*. Lawrence Erlbaum, Hillsdale NJ.

Zuckerman, M. 1985. 'Biological foundations of the sensation-seeking temperament'. In J. Strelau, F. Farler and A. Gale (eds), *The Biological Bases of Personality and Behaviour*. Hemisphere, Washington DC.

INDEX